Animal Edutainment in a Neoliberal Era

(RE)THINKING ENVIRONMENTAL EDUCATION

Constance Russell and Justin Dillon
General Editors

Vol. 15

The [Re]Thinking Environmental Education series
is part of the Peter Lang Education list.
Every volume is peer reviewed and meets
the highest quality standards for content and production.

PETER LANG
New York • Bern • Berlin
Brussels • Vienna • Oxford • Warsaw

Teresa Lloro

Animal Edutainment in a Neoliberal Era

Politics, Pedagogy, and Practice in the Contemporary Aquarium

PETER LANG
New York • Bern • Berlin
Brussels • Vienna • Oxford • Warsaw

Library of Congress Cataloging-in-Publication Control Number: 2019053312

Bibliographic information published by **Die Deutsche Nationalbibliothek.**
Die Deutsche Nationalbibliothek lists this publication in the "Deutsche Nationalbibliografie"; detailed bibliographic data are available on the Internet at http://dnb.d-nb.de/.

ISSN 1949-0747
ISBN 978-1-4331-4721-0 (hardcover)
ISBN 978-1-4331-7414-8 (ebook pdf)
ISBN 978-1-4331-7415-5 (epub)
ISBN 978-1-4331-7416-2 (mobi)
DOI 10.3726/b16172

80 Broad Street, 5th floor, New York, NY 10004
www.peterlang.com

For my father, Arturo, who taught me how to love animals.

Contents

List of Illustrations

Figures

Table

Foreword by Lori Gruen

There is no place on earth that remains untarnished by human activity. As our reach widens, our relationships with the more-than-human world tighten, becoming more strained and destructive. In the so-called Anthropocene humans are entangled in a complex web of connections with other animals and the planet, but these connections often are not affectively experienced; indeed, most of these relationships are mediated through economic markets and leave us feeling detached from nature.[1] In recognition of this disconnection, modern zoos and aquariums promote themselves as sites to reconnect, learn about other animals, and marvel at the natural world.

No longer solely designed to amuse and amaze visitors by presenting animals as fantastical or frightening spectacles, zoos and aquariums are now "edutainment" facilities, paradoxical sites where technologies and animals come together to educate people, largely young people, about the wonders of the natural world. The inconsistencies of edutainment are deftly explored in Teresa Lloro's engagement with the animals, both human and nonhuman, at the Aquarium of the Pacific. The Aquarium's mission is: "To instill a sense of wonder, respect and stewardship for the Pacific Ocean, its inhabitants and ecosystems" in an "aquarium dedicated to conserving and building Natural Capital (Nature and Nature's services) by building Social Capital (the interactions between and among peoples)." The solution to the alienation of our market disconnections to nature that the Aquarium of the Pacific proposes, then, is capital-mediated encounters with animals whose very lives are jeopardized by market forces.

Teresa Lloro argues that zoos and aquariums "biopolitically regulate and commodify animal bodies in a multiplicity of neoliberal ways" by offering visitors who spend a bit more than the admission price direct access to animal bodies. These premium "animal encounters"—like touching sharks in the Shark Lagoon or feeding Australian Lorikeets in the 5,400 square foot aviary—are designed to bring visitors closer to animals in a highly curated encounter. The goal of these orchestrated connections is to help move people to change their personal behaviors, perhaps by purchasing sustainable seafood. But can market-based edutainment really help solve the problems other animals encounter as humans carry on with their earth-destroying practices? It seems highly unlikely.

The way animals are presented in these close encounters as well as in their exhibits further distorts our relationships with them. Part of the problem is that zoos and aquariums are not places in which animals can be seen as dignified, as worthy of respect. They are valuable only insofar as they serve us or serve as ambassadors of their species. The enclosures and the scripted and unscripted "shows" are designed to satisfy human interests and desires. And the experience creates a relationship in which the observer, even a child, is free to feel dominant distance over those being observed. Writing about zoos, Ralph Acampora (2005) observes, "The very structure of the human-animal encounter is disrupted, and the interaction that is sought—encountering the animals—becomes an impossibility as the 'real' animals disappear and the conditions for seeing are undermined" (p. 71).

The sorts of shows that Teresa Lloro discusses help further inform an ethics of sight that I have been thinking about, particularly as such ethics illustrate the difficulty of respecting animal others in captivity (see Gruen, 2014). The relationship of watched and watcher under conditions of captivity, whether or not the watcher is the one who confines and controls the watched, strains the possibility of respect. Here, the biopolitical management of the lorikeets, and the human visitors, clearly illustrates the strain. The watcher and the watched are themselves surveilled by staff who try to regulate visitors' quick movements, as well as the biting and shitting lorikeets. In addition to undermining the possibility of regarding animals as worthy of respect, animals do not often have a choice to escape the human gaze. In the relative freedom of their native habitats, sharks and lorikeets, for example, can be seen or not and can watch humans, either seen or unseen by us. But in captivity, animals are forced to stop doing the things that make them indecent according to human norms and made to do things that they do not ordinarily do because humans want them to. When they try to assert their will by failing to participate, they are disciplined. What visitors end up seeing is what the curating staff want them to see.

Should the shows be stopped then? Should zoos and aquariums be shut down? Teresa Lloro does not think that abolishing zoos and aquariums is ultimately the answer, at least in the short term. And to a certain extent I agree. Even if no more animals are taken from the wild and the practice of breeding to sustain captive populations was abandoned, which would drastically reduce the zoo and aquarium populations, there will still be many long-lived animals in captivity for quite some time. But for some species, like elephants, dolphins, and whales, continued captivity is not good as there simply is not enough room in a zoo or aquarium to satisfy their need to travel long distances. We need to develop alternatives.

The people who work at the Aquarium have a "deep passion for helping animals and/or the environment and they genuinely believed that the Aquarium provided them a place to embody that passion by improving the lives of animals there." This has been my experience with the many zoo professionals I have worked with too. They are committed to relationships of respect for the animals they care for, but the environment in which to develop respect may, by its very structure, preclude that possibility. So what might be done?

Like Teresa Lloro, I think that helping zoos and aquariums embrace a sanctuary ethos will allow for a shift from the structure of neoliberal instrumentalization of captive animals toward relationships of respect. One important change would be to develop an ethics of sight that works to achieve visual equality, to see animals for who they are, as worthy of respect, and to allow ourselves to be seen by them. This also means allowing animals *not* to be seen and also ending manipulated interactions like those described in this book. Instead of touching animals, interactive educational programming could be developed that does not turn the shark into a tactile spectacle. Helping visitors see the world through a shark or a fish or a bird's eyes can help them recognize the limits of our ways of seeing the world. In being open to learning how others see, we can also expand our own perception.

"Sanctuary" is more than a word. Of course there are places that call themselves "sanctuaries" that do not put the interests, choices, and wellbeing of the animals first. These are not true sanctuaries. Some parts of some zoos operate very much like sanctuaries now, treating animals with great care and respect and allowing them to decide whether they want to be seen by the public or not.

For zoos and aquariums truly to become more like sanctuaries, in addition to providing the basic needs for wellbeing and enough space, animals must be treated with respect. Respecting captive animals at the very least involves providing them with opportunities for choice about who to spend time with, including captors and observers, and crucially, captives must be provided with the ability to escape the gaze of others. Animals should be provided with places to hide that are recognizable to them as hiding places given their species-specific behaviors. In captivity, we can respect animals by allowing them to be seen only when they wish to be seen and recognize that their lives are theirs to live without our judgments or interference. Since there is no place on earth that is beyond our reach, beyond our capacity to "make live" and to make die, rethinking our vexed relationships with others is crucial. Recognizing the limits of our ways of seeing and making others visible; shifting our relationships from interference, control, capture, and violence to empathy and care; developing humility when the future of our planet seems hard to fathom, will not happen in an average 2.5-hour zoo or aquarium visit. We need to expand our imaginations. Perhaps zoos and aquariums can play a role in that expansion, but as is clear in the book that follows, not in their current neoliberal, edutaining form.

Lori Gruen, William Griffin Professor of Philosophy and Professor of Feminist, Gender, and Sexuality Studies and Science in Society at Wesleyan University

Note

1 I say "so-called" as not everyone has accepted the idea that we are in a new geological epoch. Chris Cuomo has challenged the idea on ethical grounds.

References

Acampora, R. (2005). Zoos and eyes: Contesting captivity and seeking successor practices. *Society & Animals, 13*(1), 69–88.

Cuomo, C. (2017, September 20). The 'Anthropocene': Foregone or premature conclusion? *Earth: The Science Behind the Headlines*. Retrieved from https://www.earthmagazine.org/article/comment-anthropocene-foregone-or-premature-conclusion-examining-ethical-implications-naming

Gruen, L. (2014) (Ed.). *The ethics of captivity*. Oxford, UK: Oxford University Press.

Acknowledgements

A dissertation project, and any book that follows, indelibly results from the hard work of many people.

First, I express deep gratitude to all of my human research participants at the Aquarium of the Pacific. Given the ethnographic nature of this project, they allowed me to glimpse life in their personal office spaces, lunchrooms, and behind-the-scenes operations; many staff also graciously permitted to me to audio-record them as they taught at exhibits. During my time there, I learned how deeply most staff care about the work they do there—work for which many of them are not financially compensated. I deeply admire their commitments to bettering the lives of animals.

The nonhuman animals residing at the Aquarium have my deepest gratitude. Due to my personal and ethical commitments, doing research in a space that holds wild animals captive was quite difficult, especially as I came to realize that researching there co-implicated me in their captivity. Therefore, I must acknowledge those nonhuman animals as research participants without whom this project would not have been possible.

I also want to acknowledge the following people who helped me during the dissertation process and later as I crafted this research into a book: my family and extended family; my friends who helped me maintain sanity during an exciting and difficult time; my dissertation committee members, Begoña Echeverria (Chair), Derick Fay, Farah Godrej, and Margaret Nash; Maria Simani, Executive Director of the California Science Project, who afforded me access to funders who supported my research; and Connie Russell and Justin Dillon, the brilliant "[Re]thinking Environmental Education" Series Editors. Connie is not only a gifted colleague and mentor, but also an incredible friend who convinced me that I could write this book—and more importantly that my research project would be important to both academics and practitioners.

The University of California, Riverside's generous Dissertation Year Fellowship and the Robert C. Calfee and Nel Little Endowed Doctoral Fellowship provided me invaluable financial support for this project. I am forever grateful for this assistance.

And finally, I would like to acknowledge the editors and anonymous reviewers of *Environmental Education Research* and the *Journal of Political Ecology*. Earlier versions of Chapters Three and Six first appeared in *Environmental Education Research* and a prior version of Chapter Four in the *Journal of Political Ecology*. Editors and reviewers both provided generous feedback that improved the chapters published here.

Chapter One

Introduction

A replica blue whale, the largest creature to ever inhabit the earth, hangs from the ceiling of the Aquarium of the Pacific's (hereafter, the Aquarium) main building, the Great Hall. This replica—a human-made representation of an awe-inspiring animal—is designed, like the Aquarium's other exhibits, to inspire visitors to become curious about the ocean and its inhabitants.[1] The Aquarium is a prime example of an "edutainment" venue wherein institutions seek to simultaneously educate and entertain. Oceans occupy a vast majority of the earth's surface area (70 percent), but because humans and ocean creatures cannot reside in each other's habitats, humans still know very little about the lively world existing there. At the Aquarium visitors are enticed to come to know the oceans, and the environment more broadly, through fleshy, auditory, and, visual encounters with both living and nonliving nonhumans, much like the contact they have with the replica blue whale when initially entering the Aquarium's doors.[2] These local pedagogical encounters, and the wider political ecological and economic factors that structure them, are the focal point of this book.

Drawing on 14 months of multispecies ethnographic fieldwork I explore how the institution, its staff, and nonhumans, through edutainment practices, produce various sorts of knowledge about the environment.[3] In assemblages of people, nonhuman animals, and things, staff utilize storytelling as a pedagogical tool to entertain while simultaneously conveying climate and conservation science to the public. In doing this, they politically deploy (Ogden, Hill, & Tanita, 2013) live animal bodies in these edutainment processes, with the hope that they can both construct a natural world worthy of saving and inspire people to act in a more conservation-minded manner. As I demonstrate, these edutainment practices are deeply shaped by contemporary neoliberal thinking, as individual, market-based behavioral changes become the measure of conservation-mindedness and animal bodies are commodified and packaged to both ensure the institution's fiscal survivability and to promote conservation of the environment more broadly, sometimes at the expense of the captive animals living at the Aquarium. That is, I explicitly show how in some instances educational practices interfere with animals' ability to live a good life, recognizing that it is often difficult to understand what constitutes a good life for other species. Given that the majority of the aquarium research literature and popular activist campaigns focus on charismatic mammals like dolphins, whales, seals, and sea lions (e.g., Desmond, 1999; Wearing & Jobberns, 2011; Higham & Neves, 2015; Warkentin, 2012; Wiener, 2015), I chose to feature less charismatic beings like lorikeets and sharks to emphasize how they, too, can be marginalized in aquarium practice. This emphasis also contributes to a growing concern in the field of environmental education with the wellbeing of uncharismatic species (e.g. Boileau & Russell, 2018; Lloro-Bidart, 2018; Nxumalo & Ketchabaw, 2017; Pacini-Ketchabaw & Nxumalo, 2015).

As Claire Jean Kim (2015) notes in her text, *Dangerous Crossings: Race, Species, and Nature in a Multicultural Age*, neoliberal political philosophy certainly did not create racial or species injustice out of thin air, as colonial and imperialist endeavors manufactured these oppressions centuries ago (also see Deckha, 2012). However, neoliberalism re-inscribes and repackages these forms of injustice in new and often poorly understood ways, making it critical to further explore neoliberalism's various permutations in environmental education. These investigations are especially important in modern zoos and aquariums, which have yet to shed their colonialist and imperialist roots (Hanson, 2002; Lukasik, 2013).

Background

The Aquarium is a non-profit 501 3(c) institution in Long Beach, California dedicated to ocean conservation education, which it enacts at exhibit sites through enrolling living and nonliving creatures in teaching and learning processes.[4] Although this institution is unique given its placement in Southern California, many animal species exhibited there are also on display at institutions all over North America, making the findings of this research relevant for other institutions. In conducting pre-fieldwork at the space, which aided in the conceptualization of this study, I became curious about the kinds of environmental knowledge various Aquarium actors (including nonhumans) produced. In particular, drawing on work in science and technology studies, I wanted to understand how science gets leveraged in teaching practice in order to convince visitors that certain behaviors (e.g., consuming sustainable seafood) are more desirable than others (e.g., consuming any seafood or not consuming seafood at all).

Initially, utilizing Bruno Latour's (1987) actor-network theory (ANT), I also intended to examine how Aquarium experiences impacted both staff and visitor identity development, as they became enrolled in networks of knowledge production (or excluded from these networks). After the first few months of fieldwork, however, I realized that ethnographically studying visitor identity was methodologically challenging due to visitor transiency and my lack of ability to establish rapport with them in brief encounters. More importantly, while I found ANT useful from a theoretical standpoint due to its inclusion of the nonhuman world as actors engaged in the knowledge production process, I also became dissatisfied with ANT because it lacked a solid theoretical connection between knowledge and power. Further, while animals could certainly be included in ANT's theoretical ensemble as "knowledge makers," due to the commitment to the concept of symmetry, the data I had collected demonstrated that the relationships between humans and animals at the Aquarium were by no means "symmetrical."[5] As a result, I turned to the work of Michel Foucault who asserts that knowledge is never neutral. I aimed to better understand how and what kinds of knowledge are produced and, also following Foucault, how this knowledge makes particular subjectivities available (and others unavailable) to both visitors and staff.[6] Simultaneously, I realized that there were many non-transient research participants *residing* at the Aquarium whom I had not fully engaged because of my disciplinary preoccupation with human action. It quickly became apparent to me that engagement with the captive nonhuman animals should be a critical part of the research process.

As an animal advocate and activist since childhood, one would have thought that I would have *first* considered the animals as participants in my study. However, years of training in social theory had taught me that natural scientists studied animals and the natural world and social scientists only considered the realm of human cultures and societies—as if these things are indeed as separate as we Western cultures have been led to believe they are. (See Barbara Noske, 1997 for a discussion of this separation in research). Further, I was trepidatious given I was still somewhat mired in the post-positivist research traditions that dominate educational research (and so much of contemporary Western thought) and worried that my own positionality as an animal advocate and activist might unduly influence the research process. After engaging with feminist and critical multispecies scholarship, however, I came to realize that my background meant that I was, in many ways, *more* poised to do this important work than someone who lacked such experiences.

Given that my discipline, education, is entrenched in anthropocentric and humanist pedagogical and research paradigms (Lloro-Bidart, 2015a, Lloro-Bidart 2015b; Pedersen, 2010; Russell, 2005; Snaza, 2015), the vast majority of the educational research literature studying teaching and learning in zoos and aquariums, which is commonly characterized as "informal science education," overwhelmingly conceptualizes nonhuman living beings as objects to be utilized in human learning processes. (For exceptions see Fawcett, 2013; Lloro-Bidart 2014, 2017a,

2017b; Lukasik, 2013; Russell & Hodson, 2002). I found this troubling given that animal-focused education in environmental education has a rich, yet marginalized tradition (Russell & Fawcett, 2013; Spannring, 2016) of conceptualizing animals as participants in diverse "communities of knowers" (Fawcett, 2005, p. 276).[7] I sought, therefore, to better understand how animals are active and voiced participants (Corman & Vandrocová, 2014) in edutainment assemblages at sites like the Aquarium, yet also wanted to remain deeply attuned to how power structures, particularly those that are shaped by neoliberal thinking, limit and shape the expression of animal agency in such contexts. As a result, I both theorized the human-animal relationships formed in this space as highly local and situated, i.e., the smallest unit of analysis in human-animal relationships being "*the relation*" (Haraway, 2003), but also realized that the political economy and geographical location of the institution—a non-profit acting like a private enterprise in California's entertainment markets—radically impacts these relations. I subsequently engaged in a political ecological analysis of the site to better understand this complex interplay between macro political economic structures, like neoliberal capitalism, and local human-animal encounters. Before sketching the political ecological theoretical frameworks informing this research, I provide a brief history of zoos and aquariums to provide the necessary context required to understand the current zoo and aquarium model, which is one focused on conservation, education, and entertainment.[8]

History of Western Zoos and Aquariums

Aquariums, nature parks, and zoos emerged in the mid to late 19th century as dissatisfaction with urban life fomented and people began to both crave contact with a tamed, entertaining, and leisurely nature that could be easily accessed within or near city bounds (Hanson, 2002) and preserve what was perceived as a diminishing wilderness outside of the city (Cronon, 1996). Early zoological gardens, profoundly influenced by the vision of park planner Frederick Law Olmsted, also reflected a White middle class ethos and helped to shape what was later constructed as a proper balance of culture and nature as social reformers sought to "contain the threat posed by urbanization to moral and social order" (Hanson, 2002, p. 16). These institutions are also, therefore, deeply imbued with ideas about race and class, which Lisa Udin (2015) recently explored in her book *Zoo Renewal*. Thus, zoos and aquariums tend to offer mostly suburban and urban residents an "in-fake-situ" experience with nature characterized by a desire to "bring the outside in, or us into the outside" (Desmond, 1999, p. 177), while nature parks and protected areas provide an "in-situ" experience defined by authenticity, realism, and immersion in what is constructed as natural (Desmond, 1999, p. 186).

Early zoological parks tended to leverage captive wildlife to advance science, conserve animal species, educate, and entertain, although their aims and goals have evolved over time, especially given the contemporary pressures to preserve threatened and endangered species (Braverman, 2013; Hanson, 2002). Rather early in the formation of the North American zoological park, zoo professionals sought to distinguish these spaces from traveling menageries and circuses, which were viewed as flamboyant and commercial rather than genteel, despite the fact that both built their "animal collections" through similar means (Hanson, 2002). Contemporary American Zoological Association standards, for example, mandate that member institutions emphasize both conservation and education in their mission statements. Yet in many cases entertainment retains a prominent role in actual practice (Desmond, 1999; Lloro-Bidart, 2014, 2017a 2017b; Lloro-Bidart & Russell, 2017; Rutherford, 2012), in some cases due to a need to generate revenue.

The tension between economic and educational mandates manifests in edutainment ventures like aquariums and zoos, often impacting interpretation in complex ways. Further, although informal learning sites such as aquariums, nature parks, and zoos do not face the pressures of standardization and testing that influence pedagogical content and approach in formal educational spaces, they often remain highly conservative. Many adopt transmission modes of pedagogy, telling visitors what and who is important, frequently emphasizing scientific facts over advocacy (e.g., Russell & Hodson, 2002). And finally, aquariums and zoos typically assume that any experience with nature or other animals automatically leads to knowledge, caring, and advocacy as they embrace the contentious "nature-deficit disorder" rhetoric of writers like Richard Louv (e.g., Fletcher, 2016; Nxumalo, 2018). Although empirical research does not support this simplistic linear model of experiential environmental learning, it has persisted in formal, informal, and nonformal educational spaces (Russell, 1999).

Political Ecologies of Education

Introduction to Political Ecology

According to Mara Goldman and Matthew Turner (2011), political ecology is a cross-disciplinary field studying the "the use and control of natural resources" and "environmental change and its representations" (2011, p. 6).[9] Political ecology challenges apolitical approaches to studying human-nature relationships that dominated the 19th and first half of the 20th century (Paulson, Gezon, & Watts, 2005) by working to more closely link the social and natural/physical sciences and to focus on "the social relations of power and the formation and functioning of ecologies and landscapes" (Rocheleau, 1999, p. 22). Influenced by the field of cultural ecology, early political ecology was critiqued as functionalist (Paulson et al., 2005) and reductionist (Robbins, 2012) for its emphasis on structure and systems (Biersack & Greenberg, 2006; Robbins, 2012).[10]

As postmodern, poststructuralist, and postcolonial work influenced political ecology, much as it did environmental education, the field gradually began to focus on symbols and agency (Biersack, 2006). Political ecologists now draw on "common property theory, green materialism, peasant studies, feminist development studies, discourse theory, critical environmental history, postcolonial theory, actor-network theory" (Robbins, 2012, p. 51) and other approaches, all of which have also influenced strands of environmental education. Political ecologists, in response to earlier structural writings (Biersack, 2006), also began to question some of the fundamental humanist assumptions of the field through conversing with the environmental posthumanities (e.g., Haraway, 1988, 1991) and animal geography (e.g., Wolch &

Emel, 1998). Although postmodernism, poststructuralism, and postcolonialism similarly and significantly influenced education (McKenzie, 2005), this book is significant because it challenges the field to question its own anthropocentrism and humanism (e.g., Fawcett, 2013; Lloro-Bidart, 2015; Pedersen, 2010b; Russell, 2005; Snaza, 2015).

While traditional political ecology examines the complex relations between humans and the physical (or natural) environment, with a focus on the role of power and knowledge (Paulson et al., 2005; Goldman & Turner, 2011) the field has only minimally engaged with these complex human-nature interactions in the context of environmental education. As a nascent and interdisciplinary strand of research in education, the political ecology of education links insights from political ecology and the political economy of education to unpack human-nature relations in educational spaces, broadly defined.[11] The political ecology of education thus "explicitly explores how environmental education pedagogies, policies, and practices—in all their variety—intersect in fraught ways with gender, histories of colonialism, and political and economic currents to affect humans, other species, air, land, and water, environmental knowledge systems, and the ecosystems they inhabit" (Meek & Lloro-Bidart, 2017, p. 215). While a handful of political ecologists have begun to turn their attention to informal and nonformal environmental education sites like museums, ecotourism spaces, and animal parks/zoos (Fletcher, 2014; Rutherford 2011), they tend to theorize these sites as explicitly *political*, but not necessarily as educational. Political ecology, if theorized in relation to education, has the potential to illuminate how political economy deeply shapes institutional and teaching practice (Fletcher, 2015; Lloro-Bidart, 2014, 2017a, 2017b; Lloro-Bidart & Russell, 2017; Meek 2015a, 2015b, 2015c) and also might provide lenses to re-imagine environmental education as more just for the nonhuman animals we enroll in our teaching and learning processes.

In the context of the study that informs this book, the political ecology of education achieves several important analytic and theoretical tasks, including understanding how various scalar processes acting at the global, national, regional, and local level impact both environmental knowledge production through teaching and learning and the lives of animals enrolled in educational endeavors at the Aquarium. Further, drawing on feminist posthumanist and ecofeminist research in environmental education and beyond, the political ecology of education provides a framework to re-imagine these encounters as less exploitive. In the following sections, I outline the various theoretical frameworks that inform the three political ecologies of education I develop in this book: (1) Foucauldian and neo-Foucauldian governmentality; (2) biopolitics and biopower; and (3) feminist posthumanism.

Part II: Foucault and Neo-Foucauldian (Neoliberal) Governmentality

As Thomas Lemke (2001) explains in an analysis of Foucault's lectures on neoliberal governmentality, Foucault describes governmentality as having two sides: (1) the "art of government," where government simultaneously allows a problem to be addressed and provides solutions or particular acts to solve the problem, structuring interventions along the way and (2) the "conduct of conduct," which variously involves governing oneself and/or governing others (Lemke, 2001, p. 191). Neoliberalism differs from classical liberalism in the way that it redefines the relation between the state and the economy and due to its "encoding of the social domain as a form of the economic domain" (Lemke, 2001, p. 200). In this model, classical liberalism is inverted so that markets control states and individuals, resulting in subjects who come to be conceptualized as enterprising, rational, calculating consumers in every facet of their existence (Brown, 2003; Lemke, 2001; Rose, 1996), including their socioecological lives. The market is also viewed as a tool, and in some cases the primary tool, by which to effectuate social change, so neoliberal policies posit that wherever there is not a market or where one cannot exist, the goal is to incite people to behave as though there were a market. What at first glance seems to be less government, then, is actually illusory, as the "free" subjects of neoliberal governmentality are "controlled *through* their freedom" (Brown 2003, n.p., italics in original). And finally, although markets control subjects and the state, individuals as perceived "free" subjects remain solely responsible for the consequences of their decision-making, as research on neoliberal food policy and the alleged obesity crisis (Guthman & DuPuis, 2006) clearly demonstrates.

Green Governmentality and (Neoliberal/Disciplinary) Environmentality: Perspectives from Political Ecology

In foundational work linking Foucault's writings about governmentality to environmental issues, a task Foucault himself did not explicitly take up (Darier, 1999), Timothy Luke describes green governmentality as "the disciplinary articulations of sustainability and development [that] centre on establishing and enforcing 'the right disposition of things' between humans and their environment" (1999, p. 146) and environmentality as the embedding of "instrumental rationalities in the policing of ecological spaces" (1995, p. 66). In this way, the state and markets influencing the state establish and enforce the proper kind of relationship humans are to have with the environment, broadly construed (Luke, 1999). Luke highlights in his empirical analysis of the Worldwatch Institute, for example, that the politicized nature of modern economies and the embeddedness of the environment in these economies results in environmentalities that manage human-nature relationships in order to "realize tomorrow's environmentally sustainable economy" (1995, p. 75). Foucault's arts of government, according to Luke (1995), not only structure problems (e.g., sustainability), but also interventions and solutions (e.g., sustainable seafood and adopt-an-animal programs, like at the Aquarium).

More recently, building on this foundational thinking, political ecologists like Arun Agrawal (2005), Rob Fletcher (2010, 2015, 2017), and Stephanie Rutherford (2011, 2018) have further empirically explored and theorized environmentality and green governmentality (See also Laforge & McLachlan, 2018). In this research, environmentality tends to focus theoretical attention on environmental subject formation (Agrawal, 2005), whereas green governmentality more broadly considers the environment as "a biophysical reality, but also a site of power, where truths are made, circulated, and remade" (Rutherford, 2011, p. xvii) as the state manages and legitimates itself either through the market (neoliberally) or through the inculcation of ethical norms (disciplinarily) (Fletcher, 2010). Agrawal's (2005) research demonstrates how statistics as official knowledges and the devolution of authority to local centers, both as technologies of government, complexly worked to produce new environmental subjects in Kumaon, India's forests—subjects who are "protectors" of the forests. Fletcher (2010), in an analytically productive move with implications for understanding educational spaces, calls Agrawal's approach disciplinary environmentality—as opposed to neoliberal environmentality. Drawing on *Discipline and Punish*, Foucault's (1995) classic work that uses the "Panopticon" prison model as a heuristic to explain the operation of power in society, Fletcher (2010) shows how disciplinary environmentality is, in a sense, an internal mandate and neoliberal environmentality, an external one. For example, neoliberal environmentality's framing of ecotourism as an "incentive for conservation" works with the market to encourage cost/benefit analyses resulting in desired behaviors and to ensure economic growth as a tactic for social change (Fletcher, 2010). In contrast, disciplinary environmentality tends to eschew market approaches per se, and instead governs through normalizing processes.

In his research most explicitly framed as environmental education, Fletcher (2015) points out how educational spaces (in this case ecotourism sites) traditionally relied on the tactics of disciplinary governmentality/environmentality, insofar as they sought to inculcate subjects with particular ethical norms resulting in environmentally sound behaviors or actions not necessarily tied to market justifications. Fletcher (2015) emphasizes, however, that the reliance on disciplinary environmentalities in ecotourism appears to be increasingly giving way to neoliberal environmentalities, such as the emphasis on ecosystem services in teaching presentations. This relatively new shift in pedagogical practices at ecotourism sites suggests that other educational spaces, like aquariums, might similarly be undergoing such a move. This book, therefore, also explores this shift in the context of the Aquarium's Blue Cavern Show and Seafood for the Future Program, which are pedagogically linked and tend to narrowly frame human-animal relationships for visitors.

A Political Ecology of Education: Neoliberal/Disciplinary Green Governmentality and Environmentality

Coalescing neoliberal/disciplinary green governmentality and environmentality approaches in traditional political ecology (Agrawal,

2005; Fletcher, 2010, 2015; Rutherford, 2011) with an explicit focus on institutional and teaching practice, the first part of my analysis broadly considers the following questions, each with an explicit focus on how they frame human-animal relationships:

- Visibility: What particular problems are being illuminated and what is being obscured?
- Technical aspects: What instruments and procedures are being used to address the above?
- Formation of identities: What forms of self are presupposed by these practices of governance and what transformations are sought? (Hobson, 2013)

To answer these questions, I explore how the Aquarium's funding structure, placement within Southern California's entertainment markets, and commitment to ocean conservation education complexly interact to shape its mission and vision, which in turn influence on-the-ground practice. In this way, the mission and vision guide how the Aquarium visibilizes particular problems (i.e., potential shortages of seafood), chooses the instruments and procedures to solve these problems (i.e., storytelling at the exhibit site through the Aquarium's Seafood for the Future program that advocates the consumption of sustainable seafood thereby normalizing seafood consumption), and works to form identities (i.e., consumers of sustainable seafood who believe they are environmentally responsible). Next, using neoliberal and disciplinary environmentality lenses (Fletcher 2010, 2015), I specifically show how the Aquarium visibilizes the practice of consuming seafood through the technology of storytelling at the exhibit site and through the market-based tactics of the Seafood for the Future program. I also focus on how the Aquarium envisions environmentally responsible action for its visitors as I consider the forms of self and human-animal relationships these stories presuppose (Hobson, 2013). Such practices are governmental to the extent that they aim, in complexly layered ways, to have visitors self-regulate when they make choices about what to eat—choices that have impacts on the lives of animals.

Part III: Foucault's Biopolitics/Biopower and Neo-Foucauldian Applications to Animal Bodies

After several months of interviews with staff working closely with live animals and observations of exhibits featuring live animals available for body-to-body contact, I realized I needed an analytic framework to allow me to understand how animal bodies at the Aquarium are enrolled in and also complicate human educational experiences. The second prong of the book's political ecology education analysis uses, therefore, Foucault's ideas of biopolitics and biopower and the work of scholars who have applied his ideas to nonhuman bodies (Chrulew, 2011; Luke, 1999, 2000a, 2000b; Rutherford, 2011; Also see Cavanagh, 2018 for a recent review). Foucault (1978, 2003) developed the conceptual tools of biopolitics and biopower in order to explain how, in the nineteenth century, the "state" began to control the [human] biological in order to "make live and let die," whereas the "sovereign" had previously operated through the mechanism of "let live and make die" (2003, p. 241). Essentially, Foucault argues that both state and non-state entities made the management of life a political issue necessitating state intervention in order to foster life in particular ways. He distinguishes between disciplinary technologies (those acting on the individual body) and regulatory technologies (those operating at the level of the population). According to Kevin Grove, biopower and biopolitics can be theoretically and empirically distinguished because the former refers to "a mode of power that takes life as its object" and the latter the "techniques and rationalities of power mobilized in the pursuit of the security, growth, and development of individual and collective life" (2014, p. 22)

In the last two decades, scholars have applied Foucault's ideas about biopolitics and biopower to the management of nonhuman bodies in various spaces including zoos, theme parks, ecotourism sites, confined animal feeding operations, and botanical gardens (e.g., Chrulew, 2011; Collard, 2012a, 2012b; Luke, 2000a, b; Rutherford, 2011; Shukin, 2009; Srinivasan 2017; Stănescu, 2017; also see Braverman (2013) on the role of Foucault's pastoral power at zoos and aquariums).[12] In my research, I demonstrate that at the Aquarium, animal bodies are biopolitically assembled as they become enrolled in encounters explicitly connected to the Aquarium's neoliberal capitalist edutainment project. Because the Aquarium promotes and actively encourages human-animal inter-activity, the institution utilizes site-specific disciplinary and regulatory technologies to make animal bodies fit for these commodified visual and tactual encounters. Further, these micro-populations of animal bodies—as I demonstrate through highlighting the cases of the lorikeets inhabiting the Lorikeet Forest exhibit and the sharks displayed in the shallow touch pools of the Aquarium's Shark Lagoon exhibit—serve as ambassadors for their own species, and for the environment more generally, as they are regulated at both the individual and population level in order to "make live."[13] Following Foucault (2008), this sort of "making live" subjects the animals at the Aquarium to a kind of political death, vis-à-vis a unique type of "speciesism" where neoliberalism folds animal bodies into edutainment endeavors.[14] Drawing connections to Foucault's (2008) ideas about racism, in *Society Must be Defended*, where "if you want to live the other must die" (p. 255), the Aquarium animals are isolated from members of their own species in the environment they typically inhabit so they may be *displayed* and *touched* for the benefit of the greater conservation good. In this way, animal bodies not only serve edutainment ends, but they are specifically packaged and then commodified to serve political economic ends as they—and the encounters they become part of—ensure that the institution remains fiscally viable in Southern California's competitive entertainment market.[15]

Part IV: Feminist Political Ecology: Personhood and Affective Labor

Although Foucault's ideas about biopolitics and biopower served as a useful theoretical lens to explain how the institution enrolls animal bodies in its edutainment endeavors, they lack the analytic power to understand how some staff, in conjunction with Aquarium animals, subtly resisted the Aquarium's neoliberal agenda. The third political ecology of education I develop, therefore, turns to feminist and posthumanist political ecology and theories of personhood and affective labor (or *biopower from below*) to understand this resistance and theorize a feminist posthumanist political ecology for environmental education.[16]

Kay Milton's (2002, 2005) person-based identification and feminist posthumanist political ecological perspectives drawing on affective labor (Hardt, 1999; Hardt & Negri, 2000), as invoked by feminist scholars like Neera Singh (2013, 2015), work in the mode of feminist poststructural thought as they focus on how human discursive practices not only "speak ourselves into existence" (Davies 2000, p. 62), but also speak into existence the nonhumans with whom we share our lives. Simultaneously, they focus on how the agency and actions of nonhumans contribute to and shape relationships and knowledge production (Milton, 2002, 2005; Singh, 2013). As outlined below, these theories also firmly situate the analysis within the political economic context of the institution and the greater Southern California area, making these relations, relationships, and knowledge production inherently political, ecological, and ethical (Fawcett, 2005).

While analyses that variously examine how gender, race, ability status, body size, sexual orientation, and species oppressions are linked and have value in environmental education research (see Lloro-Bidart, 2015b; Di Chiro, 2006; Lloro-Bidart & Finewood, 2018; Maina-Okori, Koushik, & Wilson, 2018; Russell & Semenko 2016), in the feminist political ecology framework I develop, I do not take an intersectional approach that centers gender per se. Rather, I use the tools of feminist political ecology to invoke "a feminist perspective …[as an] ongoing exploration and construction of a network of learners" (Rocheleau 2015, p. 57) that does not begin with gender as the primary unit of analysis. Instead, species difference and interspecies relationality are situated within the cultural and political context of neoliberalism, which explicitly responds to feminist posthumanist philosopher Maneesha Deckha's (2012) call to better recognize the importance of culture in multispecies analyses. In this case, the aim is to unpack how the cultural work of humans and nonhumans (Hartigan, 2015) actually subverts neoliberal ideologies dominating institutional and educational practice (Deckha, 2012).

The resulting nascent conceptual framework, which considers nonhuman animals as participants in a "network of learners," and species difference as a key site of neoliberal instantiation, allows me to address the persistent anthropocentrism of environmental education research and practice (Fawcett, 2013; Russell, 2005; Snaza, 2015). I, therefore, work to decouple

> the condition of 'women' and the condition of animals ...[because] feminist work on speciesism and within animal studies should be done regardless of humans and regardless of issues of sex/gender" if it is to truly address the pressing question, "what's in it for the animal? (Clark, 2012, p. 518)

Despite the lack of focus on gender specifically, my analysis nonetheless remains feminist given its focus on the role of affect, emotion, and material relations in building human-animal relationships (e.g., Adams & Gruen, 2014; Donovan, 2006; Donovan & Adams, 2007; Gruen, 2015), and remains posthuman because it emphasizes how people come to view nonhuman entities as persons, but not in the tradition of the rational, speaking subject of humanism (Snaza, 2015). My aim is certainly not to preclude other scholars from developing intersectional political ecologies of education that center gender, but rather to invite them into a polyvocal conversation that foregrounds the linkages among various oppressions.

Person-based identification. Kay Milton (2002, 2005) mounts a challenge to constructionist theories that posit the knowledge-making process as an entirely social experience. Instead, she develops a model of interaction with living and nonliving entities called person-based identification that she grounds in a consideration of both the human and nonhuman in knowledge production. While humanist in its invocation of personhood, Milton's work is posthuman insofar as she draws on deep ecology's dissolution of the divide between the "ecological self" and the "interpersonal self" to develop personhood as relational with all aspects of the environment (humans, nonhumans, and nonliving material entities). Challenging the Western modernist notion of personhood as "what something *is*, an individual," Milton emphasizes that personhood "emerges out of what something *does* in relation to others" (2002, p. 47, italics in original). That is, she grounds personhood in a relational epistemology where "objects and entities are understood in personal terms" (2002, p. 48), undoing the notion of an autonomous human subject defined in opposition to objects (that is, nonhumans and things) specifically deemed as not persons (Snaza, 2015).

Further, Milton (2002) points out that though there is facility in locating personhood in nonhuman animals who resemble us, we can come to see any living or non-living entities as "persons like me" (for example, plants, fungi, rocks) if we develop emotional relationships with those entities. Yet Milton (2002, 2005) does not: (1) consider how the condition of the nonhuman impacts the kind of personal experience a person can have in a particular environment; and (2) fully develop the degree to which humans can identify with other entities through the recognition of negative attributes or emotions.[17] Related to the first point, for example, Constance Russell and Derek Hodson (2002) and Teresa Lloro-Bidart and Russell (2017) have demonstrated that some experiences with nonhuman animals or nature can be "miseducative" when a problematic hidden curriculum communicates ideas that counter educators' intent. In the context of the second point, critical research in animal geography (Blue & Alexander, 2015; Ginn, 2014; Lloro-Bidart, 2017c; Pitt, 2018) and environmental education (Boileau & Russell, 2018; Fawcett, 2014; Lloro-Bidart, 2017a; 2018; Nxumalo & Pacini-Ketchabaw, 2017; Pacini-Ketchabaw & Nxumalo, 2015; Taylor & Pacini-Ketchabaw, 2015) shows that the development of empathic relationships with other animals is exceedingly complex and mediated by human desires and expectations. Animals who have attributes humans perceive as negative, for example, do not receive the same kind of care and compassion as those who are perceived in a more positive light.

While posthumanist scholars like Kelly Oliver (2009), Manuela Rossini (2014), and Cary Wolfe (2013) view theories of personhood like Milton's (2002, 2005) as incompatible with posthumanist thought because of their conjuring of interests and rights-based language, I maintain that Milton's personhood could be reconcilable with feminist posthumanism due not to the prominence of "a sound basis in reason" (as in animal rights philosophy's focus on sentience, moral worth, and/or self-consciousness) or "logical bases" (as in environmental philosophy's emphasis on human welfare or nonhuman moral value) (Milton, 2002, p. 95), but for the salience of emotion, empathy, and relational understandings.

Milton's theories, however, remain irreconcilable with posthumanism to the degree that her "person" is a subject "like me," i.e., human-like. To account for this incompatibility, I expand Milton's notion of personhood and argue that it should also entail an understanding that: (1) the Other (in this case nonhuman animal) is always more than what we can experience of them; and (2) our discursive and material practices both shape and limit who the Other is and can become (Latimer, 2013; Smith, 2001, 2011; Warkentin, 2010, 2012). The feminist posthumanist theory I develop, which knots together emotion, affect, materiality, and personhood as people directly experience relationships with nonhuman animals in educational spaces, thus potentially provides a way to salvage the notion of personhood in feminist posthumanist thought. Given the significance of emotion and affect in Milton's (2002) theorizing, I place her work in conversation with feminist notions of affective labor (Hardt, 1999; Singh, 2013).

Feminist political ecology and affective labor. While an extensive review of the interdisciplinary field of political ecology is outside the scope of this book, I note here that a political ecological framework is vital to the current analysis. As Meek (2015a, 2015b, 2015c) and Teresa Lloro-Bidart (2015a, 2017a, 2017b) emphasize, political ecology, with its focus on the political and scalar (global, national, regional, and local) aspects of "the social relations of power and the formation and functioning of ecologies and landscapes" (Rocheleau 1999, 22) holds great potential for education, even though thus far it has engaged with education in a strikingly limited fashion. Feminist, posthumanist, and multispecies political ecology, in particular, challenge nature-culture binaries (Biersack 2006) that frequently ignore embodied, emotional, and political ways of knowing animals—binaries that subsequently serve as the foundation for exploitative human-animal relationships in a variety of spaces, including educational ones (see Collard, 2012a, 2012b; Baynes-Rock, 2013; Hovorka, 2006; 2012; Salazar-Parreñas, 2012 for examples of multispecies political ecological analyses involving other animals). Like Singh (2013), I specifically aim to demonstrate how some affects, such as relational and emotional connections to nonhuman animals that allow us to view them as persons, challenge neoliberal capitalist projects reproducing problematic nature-culture binaries. To do this, I draw on theories that render immaterial labor as "corporeal and affective" (Hardt & Negri, 2000, p. 293) and as productive of an "immaterial good, such as service, knowledge, or communication" (Hardt & Negri; 2000, p. 290),

Michael Hardt criticizes Foucault's idea of biopower for its focus on technologies of power that "create, manage, and control populations" and instead calls for attention to the "situation for the perspective of the labor [and laborers] involved in the biopolitical production," or what he deems "biopower from below" (1999, p. 98). In this book, for example, I show how a staff member named Jennie labors "from below" with the Aquarium's birds, producing affects, subjectivities, and "life," or embodied and emotional ways of knowing, that resist some of the Aquarium's neoliberal edutainment projects (see also Lloro-Bidart, 2014, 2017a).

In a critique of Hardt and Negri's (2000) affective labor, feminist anthropologists Laura Bear, Karen Ho, Anna Tsing, and Sylvia Yanagisako argue, "The category of immaterial (affective) labor creates a "false binary" and a "hierarchy of value" regarding different forms of labor" (2015, n.p.), binaries and hierarchies that Rhea "Juno" Salazar-Parreñas (2012) challenged in her analysis of affective and custodial human-orangutan relationships in Malaysia. While such critiques of "immaterial labor," are warranted, feminist and posthumanist affective labor perspectives resist the privileging of immaterial over material labor. Singh (2013) poignantly demonstrates, for example, how the labor that Odisha, India villagers perform is both material (they gather in the forest and perform work so that it will grow and produce timber) and affective (they develop and strengthen embodied relationships, developing affective ties). In these analyses of affect, material and immaterial forms of labor are inextricably entangled—indeed inseparable—and offer "challenges to neoliberal interventions that apply the logic of the market to human-nature relations" (Singh, 2013, p. 197). To address and dismantle this material/immaterial binary, then, in subsequent sections of this book I refer to "affective labor" in the mode of feminist and posthumanist thought (Salazar-Parreñas, 2012; Singh

2013, 2015), which includes embodied labor and affective or emotional relations between two entities.

Overview of Chapters

Chapter Two provides a brief historical overview of the Aquarium, which opened in the late 1990s. The history and geographical location of the institution, as well as its unique funding structure and placement within Southern California's entertainment market, are key to explaining how neoliberalism impacts animal-focused edutainment in this space. Although the Aquarium was constructed well after the colonial and imperial endeavors of zoo founders like Carl Hagenback (Hanson, 2002), my analysis in this book demonstrates how neoliberalism repackages some of those early practices in new ways, retaining the colonial and imperialist roots of similar institutions.

Chapter Three invokes a political ecology of education framework informed by neoliberal and disciplinary environmentality to analyze institutional and teaching practice at the Aquarium. In this analysis, I demonstrate how the institution's placement within Southern California's entertainment markets and commitment to ocean conservation education all influence how the Aquarium conceptualizes itself and its work. I focus on the case of the Blue Cavern Show and the Seafood for the Future program, which work in tandem to define a problem (declining fish stocks and possible seafood shortages) and then structure a neoliberal solution through the market (sustainable seafood consumption). I conclude by discussing the implications of this research for the field of environmental education, including unpacking how neoliberalism impacts animal-focused education especially as it relates to notions of framing environmentally responsible action.

Chapter Four explores how the Aquarium politically deploys lorikeets in the institution's edutainment project. Through carefully crafted disciplinary and neoliberal biopolitical moves, the Aquarium governs lorikeet bodies and visitor-lorikeet interactions to provide visitors an opportunity to tactually interact with sanitized wild animals. In so doing, the Aquarium aims to both promote conservation education and maintain fiscal solvency by offering visitors a popular attraction.

The interactions, however, are also heavily shaped by lorikeet agency as the birds mediate contact and sometimes defecate on or bite visitors. Despite this opportunity for agency, I nonetheless demonstrate that the resulting commodified human-lorikeet interactions limit the development of empathic human-animal relationships, which I more fully explore again in Chapters Six and Seven.

Continuing with Chapter Four's analysis of how the Aquarium biopolitically regulates animal bodies, Chapter Five explores how the various exhibits that teach about sharks produce a hidden curriculum to construct what I call the "disembodied shark." A discursive and tactual construction that relies heavily on naturalistic ethics, this shark is not in fact materially present at the Aquarium or in the oceans, but is used to send a message that sharks are worthy of human care and ought to be "saved" by humans. The construction of the disembodied shark does little for the small sharks residing at the Aquarium who are subjected to a "political death" (Foucault, 2003) as they are relegated to the category of biologically inferior and are ostracized from life in the wild. Further, it's questionable whether the Aquarium's reliance on naturalistic ethics to construct a worthy, desirable shark actually induces people to care more about real sharks in the wild, particularly those who do not measure up to the Aquarium's anthropocentric norms. In the final chapter of the book, I discuss further the implications of these practices for the fields of environmental education and zoo and aquarium interpretation.

In Chapter Six, I turn my focus to the illustrative case of one Aquarium staff member, Jennie who worked intimately with the Aquarium's lorikeets and diving birds. Continuing with the concepts of biopolitics and biopower, this chapter draws on feminist posthumanist concepts of affective labor (or biopower from below) to both build a feminist posthumanist political ecology of education for human-animal relationships and to argue for a re-thinking of the visitor experience at the Aquarium. The educational implications of this case are also explored further in the final chapter of the book.

Finally, in Chapter Seven, I begin by reviewing the implications of the political ecologies of education explored in Chapters Three through Six. I then propose that zoos and aquariums might be poised to engage in a "watershed" moment of a radical reconceptualization if they had the courage to do so. I suggest that this watershed moment might take two different forms: (1) the use of pedagogical practices that involve the acknowledgement of the banal violence inflicted on captive animals; and (2) an eventual transformation of zoos and aquariums so that they embrace sanctuary models. Such reconceptualization, I argue, would allow zoos and aquariums to better meet their purported educational and conservation goals so that they contribute to rather than impede multispecies flourishing.

Notes

1. In this book, I refer to the people who visit the Aquarium as "visitors." Most staff at the Aquarium, however, embrace the institution's customer service model and refer to them as "guests."
2. Edutainment refers here to a combination of education and entertainment. Many staff members described the Aquarium in this way. Further, my fieldwork, which revealed a focus both on education and entertainment, supports this designation.
3. In using the term "nonhuman" in this text, I acknowledge that it is fraught, but find it useful shorthand. Longer terms such as "other-than-human" and "more-than-human" still center the human and "nature" is too vague in this context. My intent here is not to dialectically define other living creatures in opposition to the human, but simply to use a term more specific than "nature" and not as cumbersome as other terms.
4. In the current study I utilize the term "enroll" in the manner of actor-network theory (ANT) approaches. That is, borrowing from Callon's (1986) original work and Nespor's (1994) application of ANT to physics education at the university level, I define enrollment as establishing networks and connections and enmeshing actors (both humans and nonhumans) in these networks. While the bulk of my analysis does not utilize the language of ANT per se, I do highlight how actors form edutainment assemblages as they produce knowledge about the environment.
5. While Latour has since moved away from some of the initial theoretical precepts of ANT, "symmetry" as he initially theorized it in *Science and Action* (1987) refers to the idea that all actors (living and non-living; human and nonhuman) should be given equal treatment in the conceptualization and understanding of networks. In the context of urban environmental education, Morgan (2017) recently and similarly emphasized that approaches inspired by Latour and Haraway often lead to a loss of focus on power and a flat ontology of scale in childhood studies research, hence the need for political ecological approaches.

6. Due to visitor transiency and my limited ability to collect rich, ethnographic field data from visitors, I did not do an in-depth analysis of visitor identity development. I instead focused on the kinds of subjectivities the institution, its staff, and animals made available to visitors vis-à-vis pedagogical practices.
7. In an article on the animal turn in environmental education that foreshadowed the 2011 special issue of the *Canadian Journal of Environmental Education* on animality, the authors argue, "What we write can determine their [more-than-humans] realities: the way that we discursively frame nonhuman animals in our research and pedagogical efforts can rationalize, perpetuate, and/or challenge our relationships with them" (Oakley et al., 2010, p. 92). This thinking shaped my entire research process.
8. Throughout the rest of the book, I describe the Aquarium as focused on "conservation education," rather than environmental education or informal science education. I use this term because the Aquarium's mission and vision statements, as well as staff's stated goals during my fieldwork there align more specifically with the aims of conservation education than broader terms like environmental education or terms more narrowly focused on science education. Further, as an American Zoological Association (AZA) member institution, the Aquarium must make conservation and conservation education one of its top priorities. See https://www.aza.org/conservation.
9. Goldman and Turner (2011) draw on Blaikie (1985), Blaikie and Brookfield (1987), Watts (1983), and Escobar (1999) to synthesize this definition.
10. For a critique of functionalist analyses of education, see McDermott and Varenne (2005). For discussions of post-post approaches in EE, see McKenzie (2005) and Russell (2005) and more recently Pedersen and Pini (2016).
11. Scholars have deployed political economy lenses to understand the relationships among politics, policymaking, and educational funding but not questions related to nature-society relations.
12. For applications of Foucault's pastoral power to animal bodies in zoos, see Braverman (2013).
13. These exhibits are described more fully in Parts II and III.
14. Animal studies scholars working in a multitude of disciplinary paradigms have variously defined "speciesism," yet they all share a commonality in that the human subject is privileged in speciesist discourses. Andrezejewski, Pedersen, and Wicklund contend that it "is the name given to the presumption of human superiority over other animals and their subjection to oppression based on this belief" (2009, p. 140). Midgley describes it as "discrimination against nonhumans, thereby branding it as an offence against equality, parallel to racism, sexism, ageism, and the like" (1983, p. 65). And Wolfe, in a critique of cultural studies, contends that speciesism "involves systematic discrimination against an other based solely on a generic characteristics—in this case, species" (2003, p. 1).

15. In Chapter Two, I thoroughly describe the research site and explain in more detail how the Aquarium's location in Southern California's entertainment markets becomes important for understanding the teaching and learning that takes place there.
16. Some theorists mostly engage with feminism to do this work and others with posthumanism. A growing number of feminist scholars meld the two perspectives together in the context of animal studies (e.g., Deckha, 2012; Hovorka; 2015; Lloro-Bidart, 2016). Some feminists also work with "new materialism," which is also sometimes couched as posthumanism depending on the theorist (e.g., Hird, 2009; Oliver, 2009; Rossini, 2014).
17. Pedersen & Stănescu refer to the "animal condition" as "the actual life situation of most nonhuman animals in society and culture, as physically and emotionally experienced with its routine repertoire of violence, deprivation, desperation, agony, apathy, suffering, and death" (2012, p. ix).

References

Agrawal, A. (2005). *Environmentality: Technologies of government and the making of subjects*. Durham, NC: Duke University Press.

Adams, C. J., & Gruen, L. (2014). *Ecofeminism: Feminist intersections with other animals & the earth*. New York, NY: Bloomsbury.

Baynes-Rock, M. (2013). Life and death in the multispecies commons. *Social Science Information, 52*, 210–227.

Bear, L., Ho, K., Tsing, A., & Yanagisako, S. (2015). Gens: A feminist manifesto for the study of capitalism. Retrieved from https://culanth.org/fieldsights/652-gens-a-feminist-manifesto-for-the-study-of-capitalism.

Bell, A., & Russell, C. (2000). Beyond human, beyond words: Anthropocentrism, critical pedagogy, and the poststructuralist turn. *Canadian Journal of Education, 25*(3), 188–203.

Biersack, A. (2006). Reimagining political ecology: Culture/power/history/nature. In A. Biersack & J. B. Greenberg (Eds.). *Reimagining political ecology*. (pp. 3–42). Durham, NC: Duke University Press.

Biersack, A., & Greenberg, J. B. (Eds.). (2006). *Reimagining political ecology*. Durham, NC: Duke University Press.

Blaikie, P. (1989). Explanation and policy in land degradation and rehabilitation for developing countries. *Land Degradation and Development, 1*(1), 23–37.

Blaikie, P. (1995). Changing environments or changing views? A political ecology for developing countries. *Geography, 80*(3), 203–214.

Boileau, E., & Russell, C. (2018). Insect and human flourishing in early childhood education: Learning and crawling together. In A. Cutter-Mackenzie, K. Malone, & E. Barratt Hacking (Eds.), *Research handbook on childhoodnature: Assemblages of childhood and nature*. New York, NY: Springer. Advance of print. https://doi.org/10.1007/978-3-319-51949-4_65-1

Braverman, I. (2013). *Zooland: The institution of captivity*. Stanford, CA: Stanford University Press.

Brown, W. (2003). Neo-liberalism and the end of liberal democracy. *Theory and Event*, *7*(1). Retrieved from http://muse.jhu.edu/journals/theory_and_event/v007/7.1brown.html

Callon, M. (1986). Some elements of a sociology of translation: Domestication of the scallops and the fisherman of St. Brieuc Bay. In J. Law (Ed.), *Power, action, and belief: A new sociology of knowledge?* (pp. 196–223). London, UK: Routledge.

Cavanagh, C. J. (2018). Political ecologies of biopower: Diversity, debates, and new frontiers of inquiry. *Journal of Political Ecology*, *25*(1), 402–425.

Chrulew, M. (2011). Managing love and death at the zoo: The biopolitics of endangered species preservation. *Australian Humanities Review*, 50, 137–157.

Clark, E. (2012). "The animal" and "the feminist." *Hypatia*, *27*(3), 516–520.

Collard, R. (2012a). Cougar figures, gender, and the performances of predation. *Gender, Place, and Culture*, *19*, 518–540.

Collard, R. C. (2012b). Cougar-human entanglements and the biopolitical un/making of safe space. *Environment and Planning D: Society and Space*, *27*(3), 489–507.

Corman, L., & Vandrovcová, T. (2014). Radical humility: Toward a more holistic critical animal studies pedagogy. In A. Nocella II, J. Sorenson, K. Socha, & A. Matsuoka (Eds.), *Defining critical animal studies: An introduction to an intersectional social justice approach to animal liberation* (pp. 135–157). New York, NY: Peter Lang.

Cronon, W. (1996). The trouble with wilderness: Or, getting back to the wrong nature. *Environmental History*, *1*(1), 7–28.

Darier, E. (1999). Foucault and the environment: An introduction. In E. Darier (Ed.). *Discourses of the environment* (pp. 1–33). Malden, MA: Blackwell Publishers.

Davies, B. (2000). *A body of writing*. Walnut Creek, CA: AltaMira Press.

Deckha, M. (2012). Toward a postcolonial, posthumanist feminist theory: Centralizing race and culture in feminist work on nonhuman animals, *Hypatia*, *27*(3), 527–545.

Desmond, J. C. (1999). *Staging tourism: Bodies on display from Waikiki to Sea World*. Chicago. IL: University of Chicago Press.

Di Chiro, G. (2006). Teaching urban ecology: Environmental studies and the pedagogy of intersectionality. *Feminist Teacher*, *16*(2) 98–109.

Donovan, J. (2006). Feminism and the treatment of animals: From care to dialogue. *Signs: Journal of Women in Culture and Society* 31(2): 305–329.

Donovan, J., & Adams, C. J. (2007). *The feminist care tradition in animal ethics*. New York, NY: Columbia University Press.

Escobar, A. (1998). Whose knowledge, whose nature? Biodiversity, conservation, and the political ecology of social movements. *Journal of Political Ecology*, *5*, 53–82.

Fawcett, L. (2005). Bioregional teaching: How to climb, eat, fall, and learn from porcupines. In L. J. Muzzin & P. Tripp (Eds.), *Teaching as activism: Equity meets environmentalism* (pp. 269–280). Montreal, QC: McGill-Queen's Press.

Fawcett, L. (2013). Three degrees of separation: Accounting for naturecultures in environmental education research. In R. B. Stevenson, M. Brody, J. Dillon, & A. E. J. Wals (Eds.), *The international handbook of research on environmental education* (pp. 409–417). New York, NY: Routledge.

Fawcett, L. (2014). Kinship imaginaries: Children's stories of wild friendships, fear, and freedom. In G. Marvin & S. McHugh (Eds.), *Routledge handbook of human-animal studies* (pp. 259–274). New York, NY: Routledge.

Fletcher, R. (2010). Neoliberal environmentality: Towards a poststructuralist political ecology of the conservation debate. *Conservation and Society*, *8*(3), 171–181.

Fletcher, R. (2014). *Romancing the wild: Cultural dimensions of ecotourism*. Durham, NC: Duke University Press.

Fletcher, R. (2015). Nature is a nice place to save but I wouldn't want to live there: Environmental education and the ecotourist gaze. *Environmental Education Research*, *21*(3), 338–350.

Fletcher, R. (2017). Environmentality unbound. Multiple governmentalities in environmental politics. *Geoforum*, *85*, 311–315.

Foucault, M. (1978). *The history of sexuality volume 1: An introduction*. New York, NY: Vintage Books.

Foucault, M. (1995). *Discipline and punish*. New York, NY: Vintage Books.

Foucault, M. (2003). *Society must be defended*. New York, NY: Picador.

Foucault, M. (2008). *The birth of biopolitics: Lectures at the College de France 1978–1979* (M. Senellart, Ed., G. Burchell, Trans.) New York, NY: Palgrave MacMillan.

Ginn, F. (2014). Sticky lives: Slugs, detachment, and more-than-human ethics in the garden. *Transactions of the Institute of British Geographers*, *39*, 532–544.

Goldman, M., & Turner, M. D. (2011). Introduction. In M. J. Goldman, P. Nadasdy, & M. D. Turner (Eds.), *Knowing nature: Conversations at the intersection of political ecology and science studies* (pp. 1–23). Chicago, IL: University of Chicago Press.

Greenberg, J., & Park. T. (1994). Political ecology. *Journal of Political Ecology*, *1*(1), 1–12.

Grove, K. J. C. (2014). Biopolitics. In C. Death (Ed.), *Critical environmental politics* (pp. 22–30). New York, NY: Routledge.

Gruen, L. (2015). *Entangled empathy: An alternative ethic for our relationships with animals*. New York, NY: Lantern Books.

Guthman, J., & DuPuis, M. (2006). Embodying neoliberalism: Economy, culture, and the politics of fat. *Environment and Planning D: Society and Space*, *24*, 427–448.

Hanson, E. (2002). *Animal attractions*. Princeton, NJ: Princeton University Press.

Hanson, E. (2002). *Animal attractions: Nature on display in American zoos*. Princeton, NJ: Princeton University Press.

Haraway, D. J. (1988). Situated knowledges: The science question in feminism and the privilege of partial perspective. *Feminist Studies*, *14*(3), 575–599.

Haraway, D. J. (1991). *Simians, cyborgs, and women: The reinvention of nature*. New York, NY: Routledge.

Haraway, D. J. (2003). *The companion species manifesto: Dogs, people and significant otherness*. Chicago, IL: Prickly Paradigm Press.

Hardt, M. (1999). Affective labor. *Boundary, 26*(2), 89–100.

Hardt, M., & Negri, A. (2000). *Empire*. Cambridge, MA: Harvard University Press.

Hartigan, J. (2015). *Aesop's anthropology: A multispecies approach*. Minneapolis, MN: University of Minnesota Press.

Higham, J., & Neves, K. (2015). Whales, tourism and manifold capitalist fixes: New relationships with the driving forces of capitalism. In K. Markwell (Ed.), *Animals in tourism: Understanding diverse relationships* (pp. 109–127). London, UK: Channel View Publications.

Hird, M. (2009). Feminist engagements with matter. *Feminist Studies, 35*(2), 329–346.

Hobson, K. (2013). "On the making of the environmental citizen." *Environmental Politics, 22*(1), 56–72.

Hovorka, A. (2006). The No. 1 Ladies' Poultry Farm: A feminist political ecology of urban agriculture in Botswana. *Gender, Place and Culture, 13*(3), 207–225.

Hovorka, A. (2012). Women/chickens vs. men/cattle: Insights on gender-species intersectionality. *Geoforum, 43*, 875–884.

Hovorka, A. (2015). The gender, place, and culture Jan Monk distinguished annual lecture: Feminism and animals: Exploring interspecies relations through intersectionality, performativity, and standpoint. *Gender, Place and Culture, 22*(1), 1–19.

Kim, C. J. (2015). *Dangerous crossings: Race, species, and nature in a multicultural age*. Cambridge, MA: Cambridge University Press.

Laforge, J. M., & McLachlan, S. M. (2018). Environmentality on the Canadian prairies. Settler-farmer subjectivities and agri-environmental objects. *Antipode, 29*(2), 359–383.

Latimer, J. (2013). Being alongside: Rethinking relations among different kinds. *Theory, Culture and Society, 30*(7/8), 77–104.

Latour, B. (1987). *Science in action: How to follow scientists and engineers through society*. Cambridge: Harvard University Press.

Lemke, T. (2001). "The birth of bio-politics": Michel Foucault's lecture at the Collége de France on neoliberal governmentality. *Economy and Society, 30*(2), 190–207.

Lloro-Bidart, T. (2014). They call them 'good-luck polka dots': Disciplining bodies, bird biopower, and human-animal relationships at the Aquarium of the Pacific. *Journal of Political Ecology, 21*, 389–407.

Lloro-Bidart, T. (2015a). A political ecology of education in/for the Anthropocene. *Environment and Society: Advances in Research, 6*, 128–148.

Lloro-Bidart, T. (2015b). "Culture *as* ability": Organizing enabling educative spaces for humans *and* animals. *Canadian Journal of Environmental Education, 20*, 93–108.

Lloro-Bidart, T. (2017a). A feminist posthumanist political ecology of education for theorizing human-animal relations/relationships. *Environmental Education* Research, *23*(1), 111–130.

Lloro-Bidart, T. (2017b). Neoliberal and disciplinary environmentality and "sustainable seafood" consumption: Storying environmentally responsible action. *Environmental Education Research, 23*(8), 1182–1199.

Lloro-Bidart, T. (2017c). When 'Angelino' squirrels don't eat nuts: A feminist posthumanist politics of consumption across Southern California. *Gender, Place, & Culture: A Journal of Feminist Geography*. *24*(6), 774–793.

Lloro-Bidart, T. (2018). Cultivating affects: A feminist posthumanist analysis of invertebrate and human performativity in an urban community garden. *Emotion, Space, and Society*, *27*, 23–30.

Lloro-Bidart, T., & Finewood, M. (2018). Looking outward and inward: What feminist theory offers the environmental studies and sciences. *Journal of Environmental Studies and Sciences*, *8*(2), 141–152.

Lloro-Bidart, T. & Russell, C. (2017). Learning science in aquariums and on whale-watching boats: The political deployment of other animals. In M. P. Mueller, D. J. Tippins, & A.J. Stewart (Eds.), *Animals in science education: Ethics, curriculum, and pedagogy* (pp. 41–50). New York: Springer.

Lukasik, J. M. (2013). ZooCrit: An introduction and "the beginning." *Journal of Curriculum and Pedagogy*, *10*(1), 4–10.

Luke, T. W. (1995). On environmentality: Geo-power and eco-knowledge in the discourses of contemporary environmentalism. *Cultural Critique*, *31*, 57–81.

Luke, T. W. (1999). Environmentality as green governmentality. In E. Darier (Ed.), *Discourses of the environment* (pp. 121–151). Malden, MA: Blackwell.

Luke, T. W. (2000a). Beyond birds: Biopower and birdwatching in the world of Audubon. *Capitalism Nature Socialism*, *11*(3), 7–37.

Luke, T. W. (2000b). The Missouri Botanical Garden: Reworking biopower and flora-power. *Organization Environment*, *13*(3), 305–321.

Maina-Okori, N. M., Koushik, J. R., & Wilson, A. (2018). Reimagining intersectionality in environmental and sustainability education: A critical literature review. *The Journal of Environmental Education*, *49*(4), 286–296.

McDermott, R., & Varenne, H. (1995). Culture "as" disability. *Anthropology and Education Quarterly*, *26*(3), 324–348.

McKenzie, M. (2005). The "post-post period" and environmental education research. *Environmental Education Research*, *11*(4), 401–412.

Meek, D. (2015a). Learning as territoriality: The political ecology of education in the Brazilian Landless Workers' movement. *Journal of Peasant Studies*, *42*(6), 1179–1200.

Meek, D. (2015b). Taking research with its roots: Restructuring schools in the Brazilian landless workers' movement upon the principles of a political ecology of education. *Journal of Political Ecology*, *22*, 410–428.

Meek, D. (2015c). Towards a political ecology of education: The educational politics of scale in southern Pará, Brazil. *Environmental Education Research*, *21*(3), 447–459.

Midgley, M. (1983). *Animals and why they matter*. Athens, GA: The University of Georgia Press.

Milton, K. (2002). *Loving nature: Towards an ecology of emotion*. New York, NY: Routledge.

Milton, K. (2005). Anthropomorphism or egomorphism? The perception of non-human persons by human ones. In J. Knight (Ed.), *Animals in person: Cultural perspectives on human-animal intimacies* (pp. 255–269). New York, NY: Berg.

Morgan, J. (2017). Learning from cities: A cautionary note about urban/childhood/nature entanglements. *Environmental Education Research, 23*(10), 1369–1378.

Nespor, J. (1994). *Knowledge in motion: Space, time, and curriculum in undergraduate physics and management*. London, UK: RoutledgeFalmer.

Noske, B. (1997). *Beyond boundaries*. New York, NY: Black Rose Books.

Nxumalo, F. (2018, December 24). How climate change education is hurting the environment. *The Hill*. Retrieved from https://thehill.com/opinion/energy-environment/422720-how-climate-change-education-is-hurting-the-environment?fbclid=IwAR0BBShkffjJL1SROs0qEA0w64RhHt5LGosb6EDywsjtwEUX0ZkXfv8B3oE

Nxumalo, F., & Pacini-Ketchabaw, V. (2017). 'Staying with the trouble' in child-insect-educator common worlds. *Environmental Education Research, 23*(10), 1414–1426.

Oakley, J., Watson, G. P. L, Russell, C., Cutter-Mackenzie, A., Fawcett, L., Kuhl, G., Russell, J., van der Waal, M., & Warkentin, T. (2010). Animal encounters in environmental education research: Responding to the question of the animal. *Canadian Journal of Environmental Education, 15*, 86–102.

Ogden, L.A., Hall, B., & Tanita, K. (2013). Animals, plants, people, and things: A review of multispecies ethnography. *Environment and Society: Advances in Research, 4*, 5–24.

Oliver, K. (2009). *Animal lessons: How they teach us to be human*. New York, NY: Columbia University Press.

Pacini-Ketchabaw, V., & Nxumalo, F. (2015). Unruly raccoons and troubled educators: nature/culture divides in a childcare centre. *Environmental Humanities, 7*, 151–168.

Paulson, S., Gezon, L., & Watts, M. (2005). Politics, ecologies, genealogies. In S. Paulson & L. Gezon (Eds.), *Political ecology across spaces, scales, and social groups* (pp. 17–40). New Brunswick, NJ: Rutgers University Press.

Pedersen, H. (2010a). *Animals in schools*. West Lafayette, IN: Purdue University Press.

Pedersen, H. (2010b). Is "the posthuman" educable? On the convergence of educational philosophy, animal studies, and posthumanist theory. *Discourse: Studies in the Cultural Politics of Education, 31*(2), 237–250.

Pedersen, H., & Pini, B. (2016). Educational epistemologies and methods in a more-than-human world. *Educational Philosophy and Theory, 49*(11), 1051–1054.

Pedersen, H., & Stănescu, V. (2012). Series editors' introduction: What is "critical" about Animal studies? From the animal 'question' to the animal 'condition.'" In K. Socha (Ed.), *Women, destruction, and the avant-garde: A paradigm for animal liberation* (pp. ix–xi). Amsterdam: Rodopi

Pitt, H. (2018). Questioning care cultivated through connecting with more-than-human communities. *Social and Cultural Geography, 19*(2), 253–274.

Robbins, P. (2012). *Political ecology*. Malden, MA: Wiley Blackwell.

Rocheleau, D. (1999). Commentary on "After nature: Steps to an antiessentialist political ecology," by Arturo Escobar. *Current Anthropology, 40*, 22–23.

Rocheleau, D. (2015). A situated view of feminist political ecology from my networks, roots, and territories. In W. Harcourt & I. L. Nelson (Eds.), *Practising feminist political ecologies: Moving beyond the "green economy"* (pp. 29–66). London, UK: Zed Books.

Rose, N. (1996). Governing "advanced" liberal democracies. In A. Barry, T. Osborne, & N. Rose (Eds.), *Foucault and political reason* (pp. 37–64). Chicago, IL: The University of Chicago Press.

Rossini, M. (2014). "I am not animal! I am a human being! I … am … a man!" Is animal to human as female as to male? In E. A. Cederholm, A. Björk, K. Jennbert, & A. Lönngren (Eds.), *Exploring the animal turn* (pp. 111–124). Lund, Sweden: Lund University.

Russell, C. (1999). Problematizing nature experience in environmental education: The interrelationship of experience and story. *Journal of Experiential Education, 22*(3), 123–128, 137.

Russell, C. (2005). "Whoever does not write is written": The role of "nature" in post-post approaches to environmental education research. *Environmental Education Research, 11*(4), 433–443.

Russell, C., & Fawcett, L. (2013). Moving margins in environmental education research. In R. B. Stevenson, M. Brody, J. Dillon, & A. E. J. Wals (Eds.), *The international handbook of research on environmental education* (pp. 369–374). New York: Routledge.

Russell, C., & Hodson, D. (2002). Whale watching as critical science education? *Canadian Journal of Science and Technology Education, 2*(4), 485–504.

Russell, C., & Semenko, K. (2016). We take "cow" as a compliment: Fattening humane, environmental, and social justice education. In E. Cameron & C. Russell (Eds.), *The fat pedagogy reader: Challenging weight-based oppression through critical pedagogy* (pp. 211–220). New York, NY: Peter Lang.

Rutherford, S. (2011). *Governing the wild: Ecotours of power.* Minneapolis, MN: University of Minnesota Press.

Rutherford, S. (2018). Green governmentality. In N. Castree, M. Hulme, & J. O. Proctor (Eds.), *Companion to environmental studies* (pp. 205–208). New York, NY: Routledge.

Salazar-Parreñas, R. J. (2012). Producing affect: Transnational volunteerism in a Malaysian orangutan rehabilitation center. *American Ethnologist, 39*(4), 673–687.

Shukin, N. (2009). *Animal capital: Rendering life in biopolitical times.* Minneapolis, MN: University of Minnesota Press.

Singh, N. (2013). The affective labor of growing forests and the becoming of environmental subjects: Rethinking environmentality in Odisha, India. *Geoforum, 47*, 189–198.

Singh, N. (2015). Payments for ecosystem services and the gift paradigm: Sharing the burden and joy of environmental care. *Ecological Economics, 117*, 53–61.

Smith, M. (2001). The face of nature: Environmental ethics and the boundaries of contemporary social theory. *Current Sociology, 49*(1), 49–65.

Smith, M. (2011). *Against ecological sovereignty.* Minneapolis, MN: The University of Minnesota Press.

Snaza, N., & Weaver, J. (2015). Introduction: Education and posthumanism. In N. Snaza & J. A. Weaver (Eds.), *Posthumanism and educational research* (pp. 1–14). New York, NY: Routledge.

Spannring, R. (2016). Animals in environmental education research. *Environmental Education Research, 23*(1), 63–74.

Srinivasan, K. (2017). Conservation biopolitics and the sustainability episteme. *Environment and Planning A: Economy and Space, 49*(7), 1458–1476.

Stănescu, V. (2017). New weapons: "Humane farming," biopolitics, and the post-commodity fetish. In D. Nibert (Ed.), *Animal oppression and capitalism* (pp. 209–228). Santa Barbara, CA: Praeger.

Udin, L. (2015). *Zoo renewal: White flight and the animal ghetto.* Minneapolis, MN: University of Minnesota Press.

Warkentin, T. (2010). Interspecies etiquette: An ethics of paying attention to animals. *Ethics and the Environment, 15*(1), 101–121.

Warkentin, T. (2012). Thinking like a whale. In J. A. Smith & R. W. Mitchell (Eds.), *Experiencing animal minds* (pp. 129–141). New York, NY: Columbia University Press.

Watts, M. (1983). The poverty of theory. In K. Hewitt (Ed.), *Interpretations of calamity from the viewpoint of human ecology* (pp. 231–262). London, UK: Allen and Unwin.

Wearing, S., & Jobberns, C. (2011). Ecotourism and the commodification of wildlife. In W. Frost (Ed.), *Zoos and tourism: Conservation, education, entertainment?* (pp. 47–58). London, UK: Channel View Publications.

Wiener, C. (2015). Dolphin tourism and human perceptions: Social considerations to assessing the human-dolphin interface. In K. Markwell (Ed.), *Animals in tourism: Understanding diverse relationships* (pp. 146–162). London, UK: Channel View Publications.

Wolch, J., & Emel, J. (1998). *Animal geographies: Place, politics, and identity in the nature-culture borderlands.* New York, NY: Verso.

Wolfe, C. (2003). *Animal rites: American culture, the discourse of species, and posthumanist theory.* Chicago, IL: The University of Chicago Press.

Wolfe, C. (2013). *Before the law.* Chicago, IL: University of Chicago Press.

Chapter Two

The Making of a Regional Edutainment Venue

Introduction

The Aquarium, located in downtown Long Beach, is the largest aquarium in Southern California, the fourth largest in the United States, and hosts over 1.5 million visitors annually. During my fieldwork there, which took place from July 2012 through August 2013, the institution featured 19 major habitats and 32 exhibits spanning two floors, housed both indoors and outdoors.[1] Reflective of the institution's overarching goals of education, its mission statement is: "To instill a sense of wonder, respect and stewardship for the Pacific Ocean, its inhabitants and ecosystems" and its vision statement is: "To create an aquarium dedicated to conserving and building Natural Capital (Nature and Nature's services) by building Social Capital (the interactions between and among peoples)." The Aquarium self-describes as a place of "neutral ground" (Atkinson, Bader, & Piper, 2010) to gather "together experts and leaders from diverse backgrounds and different perspectives, such as recreational and commercial fishermen, environmentalists, industry executives, conservation biologists, artists, designers, and storytellers

... to explore alternative solutions to difficult environmental and societal issues" (Hill, 2008, p. 4). As a purported informal science education and "free-choice" learning institution, which means that learning is voluntary, nonconsequential, and responsive to the setting (Falk & Dierking, 2000), it also seeks to foster scientific literacy so that people can make informed decisions about complex environmental issues (Atkinson et al., 2010). Despite the formalized emphasis on conservation, stewardship, and education, my fieldwork revealed that entertaining visitors is also a key feature of the Aquarium's programming.

Institutional History

From the early 1900s to the 1970s the land directly adjacent to where the Aquarium now stands was a beachside amusement center called "The Pike," complete with the world-famous roller coaster, "Cyclone Racer." (See Figure 4.1). The center, featured in countless films and television shows, such as "Abbott and Costello in Hollywood" (1948) and "The Six Million Dollar Man" (1977), attracted as many as 50,000 visitors per weekend in its heyday, including many US Navy service members and their families working at and residing in the adjacent Navy facilities (Lorscheider, 2012). The Pike closed and was demolished in 1979 after a series of business closures in the area caused by myriad factors, including: (1) the establishment of the nearby Lakewood Center Mall; (2) competition with amusement parks like Disneyland and Knott's Berry Farm; and (3) City of Long Beach municipal decisions such as the installation of the Queen Mary. This closure greatly impacted the city's tourist economy. In the 1980s and continuing through the mid 1990s, the city of Long Beach experienced what has been characterized as a severe drug epidemic and skyrocketing crime rates (Lorscheider, 2012). The Long Beach Naval Station and Long Beach Naval Shipyard closed in the mid-1990s, leaving the city in a state "synonymous with the worst symptoms of urban blight and decay" (Lorscheider, 2012).

In early 1990s, city officials and other interested stakeholders began talk of revitalizing the area, including The Pike.[2] Ironically, given that Disneyland in Anaheim was partly responsible for the closure of the original center, the initial project proposal came from the Walt Disney Company. The plan, called DisneySea, would have included a sea-themed amusement park, a cruise and pleasure ship port, a shopping, dining, and hotel complex, and the "Future Research Center," a site of oceanographic research (Hill, 2008). Fiscal and legal complications, primarily associated with environmental concerns over the filling in of 250 acres of shallow harbor and widening of the 710 freeway to accommodate the large numbers of anticipated visitors (13 million per year), halted the project not long after its initial proposal. Long Beach city leaders, however, still believed development of the area would help to revitalize the entire city of Long Beach and envisioned a waterfront area, Queensway Bay, that could include shops, restaurants, and entertainment facilities, most notably the Aquarium of the Pacific (Hill, 2008).

Since other aquarium project proposals in adjacent cities were simultaneously in progress and an aquarium was viewed as a significant financial resource for the city, Long Beach officials worked to hasten the development of an aquarium by establishing a Board of Directors and identifying a developer, Kajima Urban Development.[3] Once the Board of Directors was in place, its members decided that gaining voter or philanthropic financial support would be too lengthy. The Board and the City of Long Beach, therefore, chose to issue tax-free revenue bonds to be repaid through admissions revenue once the Aquarium officially opened in June 1998 (Hill, 2008). These bonds were initially issued by the Aquarium as a non-profit 501(c)3 institution, but in early 2000s ticket sales lagged and the Aquarium had difficulty making its bond payments (Weikel, 2000a, 2000b). Given that the original bond-issuing agency (the 501(c)3 Aquarium) could not refinance the debt due to tax laws, in 2001 the Aquarium transferred the leasehold to the City of Long Beach which agreed to allow the Aquarium to continue to operate the facility. As of 2008, the Aquarium paid a minimum of just over $3.5 million per year to the City of Long Beach, making it the only institution of its kind in the U.S. required to pay large rents to its host city (Hill, 2008).[4] Typically, aquariums receive subsidies from local governments or pay no rent at all. As a result of this unique funding structure and the fact that the Aquarium does not have significant endowment funds, it is highly dependent on visitor revenues (e.g. ticket sales, programming fees) for survival.[5]

In 2005, the Aquarium's Board of Directors drafted a Master Plan that set out to frame an expansion of the Aquarium. Four expansion projects have been completed and include: "Our Watersheds: Pathway to the Pacific" exhibit (2008), "Molina Animal Care Center" (2010), "Ocean Science Center" exhibit (2011), "Pacific Collections Gift Shop" (2013) and "Pacific Visions," which opened in 2019 (Aquarium of the Pacific, 2017). A 29,000 square foot expansion, Pacific Visions cost $53 million and includes "a state-of-the-art" immersive theater, expanded exhibition and art galleries, additional live animal exhibits, and a new exhibition hall for installations, performances, and cultural programming (Pacheco, 2017).

Currently, the Aquarium is flanked on the east side by Shoreline Village, a complex of restaurants, bars, shops, and yachting and sailing venues (see Figure 4.2). The world famous ship, the Queen Mary, is docked on the southeast end of Queensway Bay. To the northeast sits Long Beach Convention and Entertainment Center; directly north, locals and tourists can enjoy The Pike at Rainbow Harbor, which currently bears little resemblance to its old namesake. This shopping and entertainment complex primarily features corporate retail chains such as California Pizza Kitchen and Outback Steakhouse, although there is a "Historic Carousel" and "Giant Wheel" designed to give it a feel reminiscent of the old Pike Amusement Center. In the fall of 2013, Long Beach city officials began talks around bringing in a new Cyclone Roller Coaster, which would require an investment of nearly $130 million, in order to draw more tourists into the Long Beach area (Bradley, 2013). As of the timing of writing, it has yet to be constructed.

Institutional Political Economy: The Neoliberal Market

In this section, I briefly describe what I refer to as the political economy of the institution, insofar as it situates the institution within neoliberal markets that influence its practices, including educational practices. Educational researchers lament that markets also plague many formal educational institutions, such as voucher programs and charter schools that marketize entire educational regimes, including the actual political, economic, and community processes forming schools (e.g., Buras, 2015; Wells, Slayton, & Scott, 2002). Although the ecotourism literature has examined the role of funding mandates in shaping ecotourist ventures and zoo branding and marketing (Burns, 2015; Carrier, 2012; Fletcher, 2014; Higham & Neves, 2015; White, 2015), environmental education generally lacks a literature base examining how the funding structures of informal learning institutions (most of which are non-profit institutions) impact educational programming more specifically. This omission is significant given the vast number of informal educational institutions providing environmental education in the U.S. and beyond, all of which are operating in specific financial contexts. Further, few scholars have specifically examined how neoliberalism shapes educational practice in informal learning institutions (Hursch, Henderson, & Greenwood, 2015).

In this book, therefore, I render visible some manifestations of neoliberalism in practice, while also considering the ways in which neoliberal and disciplinary environmentality intersect (Fletcher, 2010, 2015, 2017; Hursch et al., 2015; Larner, 2000). This section thus highlights how the Aquarium's funding structure locates it uniquely within Southern California's education and entertainment markets, resulting in institutional and teaching practices that cater to visitors in specific (neoliberal and ethical) ways.

Due to the Aquarium's early fiscal woes, the institution implemented new measures to attract additional visitors who would provide much needed funds. These included: new displays, more educational programs, increased marketing, doubling after-hours events, and additional behind-the-scenes tours. While the Aquarium is currently financially solvent and indeed now in the midst of undertaking a large expansion project, cost-saving measures and programs to attract additional visitors continue. As an example of cost-saving, two major exhibits, Shark Lagoon and the Lorikeet Forest, were converted to be run primarily by volunteer rather than paid staff sometime prior to 2012 (staff member, personal communication). A recent example of an attempt to generate additional revenue, in 2017, the Aquarium instituted "Premium Animal Encounters" for visitors desiring one-on-one contact with particular animals. For $350 per 1.5 hour visit, these programs offer an "intimate" experience with "nature" for visitors who can afford "premium" pricing, commodifying captive animals in ways that resonate, in some aspects, with animal ecotourism in "the wild" (Burns, 2015; Wearing & Jobberns, 2011).

According to the Aquarium's website, the institution generated nearly $39.5 million in revenue in 2017. Donors included the city of Long Beach ($1 million+) and the W.M. Keck Foundation ($200,000); corporate sponsors included the Southern California Honda Dealers Association, HCP, Inc., and Swedish Fish (Aquarium of the Pacific, 2017). Yet despite these donations, "Program Services," which include revenue generated from admissions, membership dues, educational programs, and sales of lorikeet nectar, comprise the bulk of the Aquarium's funding. In 2013, the final year of my fieldwork, program services constituted approximately 66.7 percent of Aquarium funding (Aquarium of the Pacific, 2014); by 2019, the most recent 990 Tax form available online) this figure had climbed to 72.8 percent. In 2019, lorikeet nectar sales, which I will discuss in more detail in Chapter Four, generated just over $659,000, a jump of nearly $250,000 since 2015 (Aquarium of the Pacific, 2020).

Thus, as a result of its unique funding structure and the fact that the Aquarium does not have significant endowment funds or other public sources of capital, it is highly dependent on cutting costs and generating visitor revenue through program services and sales of Aquarium paraphernalia for its survival. As I will illustrate in the following chapters, this financial situation impacts its educational programs in perhaps unique ways.[6] Certainly, an analysis of the institution's history and political economy sheds important light on how institutional funding structures can embroil informal educational institutions in increasingly neoliberal (and not simply capitalist) entertainment markets (Carrier, 2012). Irus Braverman's (2013) research with zoo professionals across primarily the U.S. and Canada, for example, makes it clear that many of these professionals view zoo visitors as "consumers" of nature with particular desires and tastes, which is supported by some aquarium and zoo research literature (e.g., Linke & Winter, 2011). In the case of the Aquarium, not only does the institution focus on what James Carrier (2012) calls "the power of the purse" (p. 6) by working to present the "environment" to visitors in a way that maintains its own fiscal survivability, but it specifically leverages the market as a means to "be" more environmental in the world. Further, the Aquarium maintains the more traditionally disciplinary discourses of environmental education, which include appeals to ethics and norms, rather than markets (Fletcher, 2015). In the following chapters of the book, I demonstrate how the institution and its staff internalize such mandates so that they work to govern visitor behavior through neoliberal and disciplinary tactics while simultaneously managing animal life through biopolitical means.

Notes

1. I interviewed 20 paid staff, 20 volunteer staff, and approximately 25 visitors during my fieldwork. I also did extensive participant-observation for approximately twelve months, while also audio recording staff teach at exhibit sites. It is important to distinguish between paid staff, who have formal training in marine biology, ecology, or oceanography and sometimes education, as well as work part-time or full-time. Volunteer staff members, in contrast, typically do not have formal training and only work one day per week, though there are exceptions.
2. It is important to acknowledge here that words like "revitalization" are frequently invoked when cities undergo gentrification, which can have devastating consequences on low-income residents. For discussions of gentrification in the city of Long Beach, see Johnson (2006) and Morris (2017).
3. Areas with competing projects include: Santa Barbara, Dana Point, Orange County, downtown Los Angeles, and revitalization of the Cabrillo Aquarium in San Pedro.
4. More recent figures are unavailable. The Aquarium's 990 Tax Form for 2017, which all nonprofit organizations in the U.S. must file annually, does not have a line item specifying rent expenses to the city, though "occupancy" was nearly $2.1 million.
5. Kathy Rudy similarly notes in her book, *Loving Animals*, "Even though most AZA-approved zoos are nonprofit establishments, the bottom line for most is clearly financial. In the world of late capitalism and increasing privatization, financial solvency emerges as the primary measures of success for many zoos" (2011, p. 126).
6. For comparison purposes, the 990 tax forms of other informal education institutions in the Southern California area (publicly available online) indicate reliance on "Program Services" revenue as follows, with the year of available data

in parentheses: Cabrillo Aquarium (2012), 25 percent; Descanso Gardens (2012), 23.3 percent; Natural History Museum of Los Angeles (2012), 31.9 percent; Los Angeles Zoo (2012), 18.4 percent; Pacific Marine Mammal Center (2013), 5.4 percent; San Diego Zoo (2013), 62.8 percent. Each of these institutions has a vastly different operating budget and serves a diverse number of visitors each year, but the overall trend demonstrates that the Aquarium more significantly relies on Program Service revenue than most other institutions, with the exception of the San Diego Zoo. Staff have indicated that the Aquarium competes with other corporate institutions as well, including Disneyland, Knott's Berry Farm, and Sea World.

References

Aquarium of the Pacific. (2013). *Aquarium of the Pacific 2013 annual report*. Long Beach, CA: Aquarium of the Pacific.

Aquarium of the Pacific. (2014). 990 Tax Form 2013.

Aquarium of the Pacific. (2017). *Campus master plan*. Long Beach, CA: Aquarium of the Pacific. Retrieved from http://www.aquariumofpacific.org/pacificvisions/architecture/campus_master_plan. Accessed September 1, 2020.

Aquarium of the Pacific. (2020). 990 Tax Form 2019. Retrieved from https://www.aquariumofpacific.org/downloads/Tax_Form_990_and_990-T_2019_Public_inspection_2019.pdf. Accessed December 28, 2020.

Atkinson, P., Bader, D., & Piper, H. (2010). *Aquarium of the Pacific 2010 annual report*. Long Beach, CA: Aquarium of the Pacific.

Bradley, E. (2013). Long Beach council moves forward with Cyclone Racer roller coaster proposal. *Press Telegram News*. Retrieved from http://www.presstelegram.com/general-news/20131001/long-beach-council-moves-forward-with-cyclone-racer-roller-coaster-proposal. Accessed September 1, 2020.

Braverman, I. (2013). *Zooland: The institution of captivity*. Stanford, CA: Stanford University Press.

Burns, G. L. (2015). Animals as tourism objects: Ethically refocusing relationships between tourists and wildlife. In K. Markwell (Ed.), *Animals in tourism: Understanding diverse relationships* (pp. 44–59). London, UK: Channel View Publications.

Busa, J. H., & Gardner, R. (2014). Champions of the movement or fair-weather heroes? Individualization and the (a)politics of food. *Antipode*, *47*(2), 323–341.

Carrier, J. G. (2012). Dollars making sense: Understanding nature in capitalism. *Environment and Society: Advances in Research*, *3*, 5–18.

Falk, J. H., & Dierking, L. D. (2000). *Learning from museums: Visitor experiences and the making of meaning*. Walnut Creek, CA: AltaMira.

Fletcher, R. (2010). Neoliberal environmentality: Towards a poststructuralist political ecology of the conservation debate. *Conservation and Society*, *8*(3), 171–181.

Fletcher, R. (2014). *Romancing the wild: Cultural dimensions of ecotourism*. Durham, NC: Duke University Press.

Fletcher, R. (2015). Nature is a nice place to save but I wouldn't want to live there: Environmental education and the ecotourist gaze. *Environmental Education Research, 21*(3), 338–350.

Fletcher, R. (2017). Environmentality unbound: Multiple governmentalities in environmental politics. *Geoforum, 85*, 311–315.

Higham, J., & Neves, K. (2015). Whales, tourism and manifold capitalist fixes: New relationships with the driving forces of capitalism. In K. Markwell (Ed.), *Animals in tourism: Understanding diverse relationships* (pp. 109–127). London, UK: Channel View Publications.

Hill, S. (2008). *History of the Aquarium of the Pacific: The first 10 years.* Long Beach, CA: Halbert Hargrove.

Hursch, D. W., Henderson, J. A., & Greenwood, D. (2015). Environmental education in a neoliberal climate. *Environmental Education Research, 21*(3), 299–318.

Johnson, D. (2006, July 30). Where a porn palace stood. *Los Angeles Times*. Retrieved from http://articles.latimes.com/2006/jul/30/opinion/op-johnson30. Accessed September 1, 2020.

Larner, W. (2000). Neo-liberalism: Policy, ideology, governmentality. *Studies in Political Economy, 63*, 5–25.

Linke, S., & Winter, C. (2011). Conservation, education or entertainment: What really matters to zoo visitors? In W. Frost (Ed.), *Zoos and tourism: Conservation, education, and entertainment*. Buffalo, NY: Channel View Publications.

Lorscheider, M. K. (2012). *Reinventing Long Beach: The fight for space and place in Post-Cold War Long Beach, 1990–1999*. Unpublished master's thesis, California State University Long Beach, Long Beach, California.

Morris, A. (2017, July 14). As more outside developers invest in Long Beach, how will communities maintain their identities? *Long Beach Post*. Retrieved from https://lbpost.com/life/2000011188-as-more-outside-developers-invest-in-long-beach-how-will-communities-maintain-their-identities. Accessed September 1, 2020.

Pacheco, A. (2017, February 3). Major Aquarium of the Pacific expansion breaks ground. *The Architect's Newspaper*. Retrieved from https://archpaper.com/2017/02/aquarium-of-the-pacific-expansion/#gallery-0-slide-0. Accessed September 1, 2020.

Rudy, K. (2011). *Loving animals: Toward a new animal advocacy.* Minneapolis, MN: University of Minnesota Press.

Wearing, S., & Jobberns, C. (2011). Ecotourism and the commodification of wildlife. In W. Frost (Ed.), *Zoos and tourism: Conservation, education, entertainment?* (pp. 47–58). London, UK: Channel View Publications.

Weikel, D. (2000a). Aquarium gets state aid as revenue drops. *Los Angeles Times.* Retrieved from http://articles.latimes.com/2000/jul/01/local/me-46667. Accessed September 1, 2020.

Weikel, D. (2000b). Financially troubled aquarium seeks to refinance debt. *Los Angeles Times* Retrieved from http://articles.latimes.com/2000/nov/22/local/me-55720. Accessed September 1, 2020.

Wells, A. S., Slayton, J., & Scott, J. (2002). Defining democracy in the neoliberal age: Charter school reform and educational consumption. *American Educational Research Journal, 39*(2), 337–361.

White, L. (2015). Zoos Victoria: Branding, marketing, and designing. In W. Frost (Ed.), *Zoos and tourism: Conservation, education, entertainment?* (pp. 189–203). London, UK: Channel View Publications.

Chapter Three

Neoliberal Subjectivities and the Pedagogy of "Sustainable Seafood"

Introduction

In a special issue of the journal *Environmental Education Research* devoted to the topic of neoliberalism, guest editors David Hursch, Joseph Henderson, and David Greenwood (2015) echoed the point feminist geographer Wendy Larner made at the turn of the new millennium, "If neo-liberalism cannot be reduced to a single set of philosophical principles or a unified political ideology, nor is it necessarily linked to a political apparatus, this will encourage us to think about different versions of neo-liberalism" (2000, p. 21). In suggesting that there are many manifestations of neoliberalism, Hursch et al. (2015) not only briefly sketch a history of neoliberalism in environmental education, but also invite readers to contribute to the growing body of scholarship that examines the unfolding of such "neoliberalisms" in environmental education. According to Larner's (2000) influential work, neoliberalism can be interpreted in three ways: as a policy framework, as an ideology, and through the lens of governmentality, the last of which draws heavily on Foucault and neo-Foucauldian scholars. In this chapter I take up Hursch et al.'s (2015) call, like others (Fletcher, 2010, 2015, 2017; Hobson, 2013), by invoking a governmentality lens to analyze educational programming at the Aquarium.

Although the Aquarium and its staff do not specifically use the language of neoliberalism or neoliberal governmentality, the institution views itself in a neoliberal fashion—that is, located within markets that legitimate its success—and believes that its visitors (and potential visitors) adhere to such neoliberal ideologies as engaging in cost/benefit analyses and desiring to care for the self (Brown, 2003), particularly as it relates to making environmentally responsible choices. As a result, the Aquarium caters to these visitors while simultaneously seeking to interpellate them into behaving in ways perceived as beneficial to the ocean and its inhabitants—in this case, consuming sustainable seafood. To do this, the institution and its staff rely both on ethical or disciplinary appeals (e.g., normalizing "doing the right thing") and market-based approaches or neoliberal appeals (e.g., tailoring consumptive practices).

Having already broadly established how the Aquarium's funding structure impacts the conceptualization and enactment of educational practice as edutainment in Chapter Two, I draw on Rob Fletcher's (2010, 2015, 2017) work on neoliberal and disciplinary environmentality to specifically unpack the ways in which the Aquarium relies on market and ethics-based tactics to story environmentally responsible behavior. While scholars have demonstrated that zoos and aquariums promote neoliberal ideologies like "consuming to conserve" (Braverman, 2013, p. 44), this chapter evinces how "consumptive market logic[s]" and "neoliberal notions of environmentalism" have "infus[ed] their way into our thinking" (Hursch et al., 2015, p. 308) so as to become tacitly enacted and ordinary. And finally, recognizing the increasing interest in education and food studies, particularly in regards to "food animals" (e.g., Kahn, 2011; Lloro-Bidart, 2019; Rice, 2013; Rowe, 2010, 2012, 2016), this chapter explicates how disciplinary and neoliberal governmentalities in teaching work in tandem to visibilize and then govern an intimate domestic practice—the eating of fish, mollusks, and crustaceans—through the rules of the market. Here, ocean inhabitants only have value for the ecosystem services or food they provide, which problematically reduces them objects lacking intrinsic value.

Neoliberal and Disciplinary Environmentality: Storying Environmental Action in Practice

The Aquarium's management staff encourage paid teaching staff to tell stories at the exhibit site in a manner that avoids what staff call "crisis-framing" and instead teaches visitors that they are "innovators."[1] Innovation, as I show, emerges primarily as the ability to make personal and individual behavioral changes so that visitors feel easily capable of acting more environmentally responsibly. That is, environmental responsibility is defined solely as individual action. These stories are semi-scripted and typically involve some ad-libbing as staff members include portions, but not all aspects, of the entire script. The scripts tend to present scientific facts followed by appeals for behavioral changes grounded in neoliberal and/or disciplinary justifications. These discursive moves are combined with the viewing of captive animals behind glass who interact with Aquarium divers or husbandry staff.[2] Husbandry staff typically care for animals behind the scenes, but also sometimes participate in live shows by sharing facts about animal behavior while animals actually perform these behaviors for human audiences. Aquarium divers similarly participate in shows by conversing with an interpreter on what is referred to as the "dry side" of the Aquarium. As carefully crafted discursive and material stories, they serve as the "conduct of conduct" (Lemke 2001, p. 191) by implying that people can act more sustainably by adopting new behaviors.

For example, Helen, one of the management staff members overseeing educational programming states "Science is facts and figures," and "You're not gonna move people to behavior change with facts and figures. You're gonna move them when you connect that emotion and you make them emotionally invested in whatever that is." In my interview with her and in other conversations we had, Helen frequently mentions emotion as a means to inspire behavioral change, which aligns with research in environmental education demonstrating the importance of affect and emotion in learning (e.g., Kelsey & Armstrong, 2012; Lloro-Bidart, 2018; Steele & Scott, 2016). However, Helen does not necessarily consider how the Aquarium frequently draws on neoliberal frames, such as consuming to conserve. In the following analysis, I demonstrate that despite Helen's commitments to the use of what she refers to as emotional appeals in teaching ("emotional investments"; "giving people hope"; "giving people one thing they can do"), both neoliberal (market-based) and disciplinary (ethical) frames work in tandem at the exhibit site to story environmentally desirable behavioral changes for visitors in particular, and sometimes problematic, ways.

The "Blue Cavern" Show and "Seafood for the Future": Normalizing Sustainable Seafood Consumption

To illustrate the storying of saving the environment through individual behavioral change, I draw on the Aquarium's Blue Cavern Show and its Seafood for the Future program. The Blue Cavern Exhibit, located downstairs in the large entry space of the Aquarium known as the "Great Hall," is modeled after Santa Catalina Island's Blue Cavern Point, a kelp forest ecosystem off the coast of Southern California. The exhibit, spanning three stories, contains 142,000 gallons of water and houses leopard sharks, California moray eels, California sheephead, giant sea bass, garibaldi, and other organisms. Fifteen-minute shows occur here several times per day as a paid interpreter and volunteer dive staff member dialogue. Dive staff wear special underwater microphones along with the rest of their scuba gear. Their voices are audible, but often muffled as staff have to coordinate breathing and speaking. The shows are scripted, though dry-side interpreters and dive staff typically ad lib as they highlight stories and facts they consider particularly salient or interesting. During the busy season (summer) staff give more shows per day to accommodate the increased number of visitors. As the excerpt below illustrates, staff tell a story about the recovery of two species of sea bass and then follow with a description about what it means to act sustainably: reducing overfishing *through* sustainable seafood consumption. Notice that when Justin, a part-time, paid teaching staff member on the dry side teaches about sustainability by conversing with the diver, Jerry, he uses both neoliberal (market-based) and disciplinary (ethics-based) discourses, illustrating just how complexly entangled both strategies are at the Aquarium.[3]

> **Justin:** That giant sea bass, that cast, is about seven feet long and in real life the sea bass can weigh up to 500 pounds. But to get to that big, it lives about 75 years before it actually reaches that mature size. And, in addition, because they're so big they have no natural predators. Meaning, the only predators they have that have led them to near extinction are humans, because spear fisherman who speared them when they swam up to them. Didn't they?
>
> *Jerry, the diver, explains that sea bass are easy prey for humans to hunt.*
>
> **Justin:** That's right. But like he said, they established marine protected areas where giant sea bass have the ability and safety to grow and mature. They have now come back strong in numbers. In addition, it's actually illegal to harvest any giant sea bass in their natural habitats. But, in talking about success stories, there's also a success story at the top of our exhibit. Those torpedo-looking fish, especially that giant one over there, we usually call them white sea bass. They're also a success story.
>
> *Jerry shares with the audience that white sea bass are bred in hatcheries so they're sustainable.*
>
> **Justin:** And sustainable is really important here. Because what does sustainable really mean? It essentially means that, you know, some of us like seafood, some of us love eating yellow tail tuna or any other fish out there, but we have to keep in mind how long it takes for the fish to actually reproduce, have babies, and also how long they can live for. Because if we keep taking fish out of the ocean, eventually they will all disappear and there will be no more food left for us in the ocean. And so sustainable means, so sustainability is actually important because we're keeping in mind that we're not over-taking fish, we're allowing their numbers to replenish and we're only taking a few out at a time. So sustainable seafood is an important concept. And, if you guys are interested, I have cards about our local restaurants that offer sustainable seafood.

In this happy twist on a "Tragedy of the Commons"-esque story (Hardin, 1968), visitors are taught that fish have value primarily because they are food "for us in the ocean"—an ecosystem service of sorts, although Justin does not invoke the exact language of ecosystem services. While he describes basic physiological characteristics of the giant sea bass and explains that the species have recovered thanks to the establishment of marine protected areas that legally forbid wild capture, throughout the show Justin eschews any substantive aesthetic or intrinsic justification for either of the sea bass species' value. Such locating of species' or the environment's value solely in the services they provide to humans, as Fletcher points out, resonates with neoliberalism because it justifies environmental action "in terms of cost-benefit analyses and its contribution to economic growth" (2015, p. 340). Further, Justin makes a more direct connection to markets when he suggests to visitors that they can be more sustainable through consuming sustainable seafood, offering them cards to guide them as they choose their dining experiences at local restaurants.

The act of "consuming to conserve," as Irus Braverman (2013) highlights, permeates zoos and aquariums across North America, as visitors are encouraged to purchase food to feed animals, buy trinkets in gift shops, and to pay fees to symbolically adopt endangered animals, all while learning about the necessity of changing personal consumption habits. At the Aquarium, this "consuming to conserve" simultaneously visibilizes individual food practices and enrolls the local Long Beach business community in the Aquarium's market-based approaches to practicing sustainability. The Seafood For the Future program that Justin describes in the Blue Cavern Show when he offers information about local restaurants that offer sustainable seafood, primarily works to conserve through the market. As a passage from the Aquarium's website further explicates, the Seafood For the Future program not only allows consumers to participate in "stewardship" through their purchases, but they can also enjoy "great food"—a "win-win" argument that resonates with rhetoric characteristic of the alternative food movement that suggests, "… you can buy something that you will enjoy [fish or seafood], and in so doing you will benefit some general cause, such as 'sustainability …' "(Busa & Garder, 2014, p. 9).

> At the heart of the program are the marketing resources of the Aquarium of the Pacific. Seafood for the Future (SFF) recognizes that the supply and demand for seafood can be influenced through incentives, marketing, and providing rewards for adopting better practices. Partners at every step of the supply chain have a chance to be featured at Aquarium events, festivals, dinners, and in our literature. Their business ultimately increases as more consumers look for great food from environmentally conscious businesses.
>
> Consumers have the power to influence sourcing with their wallets. SFF's [Seafood for the Future's] partner website makes it easy to find restaurants and businesses working to provide seafood from responsible sources. Special promotions are also available at participating restaurants in the form of

> discounted Aquarium admission, giving consumers an opportunity to learn about ocean stewardship at the Aquarium and become part of our community of concerned citizens and seafood lovers. (Aquarium of the Pacific, 2015)

In the spirit of neoliberal governance, the Seafood For the Future program takes a market and incentive-based approach to conservation, suggesting that individual consumers can make sustainable choices through "sourcing with their wallets." Operating under the assumption that most seafood is consumed in restaurants, and understanding that the practice of purchasing sustainable seafood at grocery stores and local markets is confusing to consumers, partly because of the sheer number and variety of eco-labels (Gutierrez & Thornton, 2014; Hallstein & Villas-Boas, 2013), the Aquarium developed Seafood For the Future. This program mainly targets local businesses directly instead of consumers, although consumers can also view the labels online prior to purchasing seafood at grocery stores and local markets: "Gold List" species ("recommended based on their resiliency"); "Specific Stocks" (consumers must know region, type of fishing gear, and fish name); "Branded Products" ("brands committed to using more sustainable methods"); and "Local Seafood" ("what's in season now"). Businesses who become Seafood For the Future "Full Partners"—what Julie Guthman (2007) would refer to as voluntary eco-labeling because businesses are not mandated by the state to do so and the labels are designed to attract, rather than stave off consumers—make commitments to purchasing sustainable seafood that is then offered in their local restaurants. To ensure compliance—since, as Guthman (2007) highlights, the market does not always function in neoliberalism without rules and rule-making bodies—Seafood For the Future staff biannually evaluate partners, who in return receive rewards from the Aquarium like free advertising and marketing in a variety of venues. Diners at these restaurants, as described in the excerpt above, can also reap rewards in the form of discounted Aquarium admission.

The Aquarium's Blue Cavern Show, in conjunction with Seafood For the Future, works to teach visitors about plunging fish populations, structuring and visibilizing a problem (Hobson, 2013; Luke, 1995) and grounding the significance of this loss in the language of ecosystem services—i.e., potential decreases in seafood supplies. Then visitors have the opportunity to concretely ameliorate this problem through simple changes in consumption habits—changes facilitated by the market-based approaches of a neoliberal instrument and solution (Hobson, 2013; Luke, 1995). Yet the language of consuming sustainable seafood, chronicled through the recovery narratives of the giant sea bass and white sea bass at the Blue Cavern Show, and interpreted in conjunction with Aquarium messaging promoting sustainable seafood, can also be interpreted disciplinarily and explicitly related to neoliberal governmentality. Fletcher points out, for example, that disciplinary tactics are "necessary to construct the rational actors upon which neoliberal governmentality would then operate" (Fletcher, 2010, p. 177). To work toward normalizing this practice so that visitors might internalize it as a social norm (Fletcher, 2010), Aquarium staff announce during nearly every semi-scripted and non-scripted show that Aquarium animals themselves consume restaurant-quality sustainable seafood. In this way, they visibilize and politicize what Kersty Hobson (2013) refers to as a "domestic practice" and stress the urgency of what Guthman (2007) calls "knowing where your food comes from" so as to structure proper interventions for that practice. That is, disciplinary governmentality works in this case not to demonstrate that the Aquarium animals are rational, calculating economic actors, but rather as a means to raise "awareness of the intimacy of food" (Guthman, 2007, p. 1175) and to subtly convince visitors that consuming sustainable seafood is something so important that the Aquarium animals do it too. This positioning is similar to what Fletcher describes as using "conservation celebrities" as "charismatic authority," (2010, p. 177) as many of the Aquarium animals consuming sustainable seafood are themselves the charismatic megafauna who draw visitors to the Aquarium.

Implications and Conclusions

In this concluding section, I draw out three broad implications of this research for the field of environmental education: (1) an understanding of the how neoliberal thought impacts not only the formation of non-profit informal education institutions, but also the kinds of programming these institutions offer; (2) an examination of the ways in which neoliberal and disciplinary environmentalities in a political ecology of education framework bring key insights to bear on institutional and teaching practice; and (3) a coalescing of the broad interdisciplinary field of food studies with environmental education.

Neoliberalism and the Fashioning of Educational Institutions

As neoliberal ideologies and governance sweep educational and socio-ecological systems across the world, formal educational institutions increasingly find themselves enmeshed in market rationalities (Buras, 2015; Hursch, et al., 2015; Jickling & Wals, 2008; McKenzie, 2012; McKenzie, Bieler, & McNeil, 2015; Wells, Slayton, & Scott, 2002) that deeply shape practice. While James Carrier (2012) and Braverman (2013) each discuss how ecotourism parks and zoos (informal education venues) view their visitors as consumers with particular desires and tastes that must be met in order to maintain fiscal survivability, neither theorizes their research through the lens of education. The current chapter's political ecological analysis contributes, then, to scholarship on zoos and aquariums more generally, as well as to similar scholarship analyzing informal educational spaces (Fletcher, 2015; Rutherford, 2011) by demonstrating how the Aquarium's funding structure, placement within Southern California's entertainment markets, and commitment to ocean conservation education complexly interact so that its mission and vision guide on the ground practice in both neoliberal and disciplinary ways. This research suggests that a political ecology of education framework has value insofar as it provides a lens for understanding how educational funding structures impact environmental education programming on the ground.

Storying Environmentally Responsible Actions: Neoliberal and Disciplinary Environmentalities

This chapter's political ecology of education framework, informed by neoliberal and disciplinary environmentality, unpacks how the traditional disciplinary tactics of environmental education (Fletcher, 2015) work in tandem with neoliberal tactics to provide visitors with an education about what sustainable seafood is and why it matters. In this analysis, I illuminate how the Aquarium visibilizes particular environmental problems through the neoliberal lenses of ecosystem services and consumption (seafood shortages; lack of sustainable seafood consumption), intervenes to address these problems (storying them at the Blue Cavern Show; managing the Seafood For the Future program), and presupposes a visitor who wants to act more sustainably—and is willing to do so by embracing individual dietary changes connected to the market. Contributing to a burgeoning environmental education literature that examines how "neoliberal notions of environmentalism can infuse their way into our thinking" (Hursch et al., 2015, p. 308), I conclude here by teasing out implications of this neoliberalism for environmental education.

Environmental studies scholars, and more recently environmental education scholars, have mounted a significant critique of individualized and market-based (i.e., neoliberal) approaches, pointing to how they define inherently collective problems, like the production and consumption of sustainable seafood—or even what sustainable seafood actually means—in terms of the individual (Foster, 2002, 2009; Guthman, 2008; Schindel Dimick, 2015; Weissman, 2015). Further, these approaches often assume a sort of unified subject who can easily access and afford food at any cost, ignoring issues of environmental and food justice (Agyeman & McEntee, 2014; Guthman, 2008; Haluza-DeLay, 2013; Mueller, 2009). At the Aquarium, this focus on the individual occurs not only because the institution and some of its staff adhere to what would be considered typical neoliberal thinking (conceptualization of a rational, calculating consumer), but also because staff believe that providing individual visitors "one thing they can do" will both create a positive Aquarium experience (that is inspire repeat visits) *and* mobilize them to act.

In the current case of promoting "ocean conservation," what results is an intricately layered endeavor that fails to ask, in the words of Eileen Crist, "that we view the timeworn stories of human ascent with the deep suspicion they deserve, see through the self-serving ontology of the world recoded as 'resources,' 'natural capital,' and 'ecological services,' and question what it is we are salvaging in desiring to sustain the human enterprise" (2013, pp. 136–137). When institutional and staff discourses intimate to visitors that creatures living in the ocean (e.g., fish, mollusks, and crustaceans) only have value because of the services (e.g., food) they provide to "the human enterprise" and then commodify that value through market-based approaches to conservation, they narrow the field of ethical relationships with animals and the environment to an amalgam of capital, services, and commodities—in short, they suggest fish have value because they are food for us. "Making people move," which management staff member Helen cited as a major goal of the Aquarium, comes to mean being a better consumer of seafood.

I did not hear Aquarium staff suggest, for example, that visitors limit or reduce their seafood consumption nor did they advocate any form vegetarianism. Although vegetarian and vegan diets are certainly not beyond reproach and clearly have multiple social and environmental consequences, the recommendation that visitors simply eat seafood differently, rather than in reduced quantities or not at all, further demonstrates how Aquarium ideologies and pedagogical practices are tied to market logics. These ideologies do not demand that visitors fundamentally reconsider their relationships with animals they consider food, but rather rely on new markets—those coded as "sustainable"—to teach visitors that they can continue to consume animals with minor modifications. Such a position is unsurprising, perhaps, given the complexities associated with teaching about veganism and vegetarianism (Kahn, 2011; Rice, 2013; Rowe, 2011, 2012, 2013; Lloro-Bidart, 2019; Oakley, 2019; Russell, 2019), particularly with respect to fish, mollusks, and crustaceans. These ocean dwellers lack the charisma of mammals like cows, pigs, and chickens and are relegated to the bottom of a human-constructed hierarchy, despite recent evidence that they feel pain and distress (Balcombe, 2016; Montgomery, 2015; Potts, 2017).

Further, while there is no extant research documenting whether the Aquarium and the Seafood For the Future program actually mobilize consumers to "source with their wallets" to the extent that they have any real impacts on fish populations, the research literature does suggest that the Aquarium's strategy of eco-labeling, more generally, is fraught (Elgert, 2012; Gutierrez & Thornton, 2014; Guthman, 2007).[4] Guthman (2007), for example, views eco-labels as a form of environmental governance and contends that they are problematic for myriad reasons, including how they essentially allow consumers (who may not be adequately informed) to "mak[e] regulatory decisions about ecological and public health risk, working conditions and remuneration, and even what sort of producers of what commodities should be favored in the world market" (p. 472). Further, building on Guthman's foundational work, Laureen Elgert (2012) shows how the labeling of sustainable soy is a thorny political process designed to "put the global consumer at ease," failing to address issues of inequitable land distribution to peasants in the countries that produce so-called sustainable soy. While the Aquarium, through its storying of environmentally responsible action at the Blue Cavern Show, seeks to address Guthman's (2007) concerns over adequate consumer knowledge to support such important decision-making, it is unlikely that after the average 2.5 hour stay, Aquarium visitors would be well-informed about all the regulatory decisions Guthman (2007) identifies. My findings then, grounded in a political ecology of education framework, suggest that environmental education researchers and educators take a more political and critical approach to researching and teaching about eco-labels.

Neoliberal and Disciplinary Environmentalities in Environmental Education and Critical Food Systems Education

Finally, while there is a literature base in critical food studies examining how various governmentalities shape food movements, practices, and labeling (Agyeman & McEntee, 2014; Elgert, 2012; Guthman, 2007), there has only been minimal engagement with this literature in environmental education (for exceptions see Allen & Guthman, 2006; Crosley 2013; Drake, 2014; Hayes-Conroy; 2010; Weismann, 2015) despite growing interest in critical food systems education (Lloro-Bidart, 2019; Meek & Tarlau, 2015; Oakley & Russell, 2015; Stapleton, Cole, Washburn, Jason, & Alvarado, 2017). In relation to fisheries/seafood education, the extant environmental education literature base is also limited, suggesting a need for further research. Several articles focus mainly on assessing the effects of various fishing education programs on behavioral changes (Kemmerly & Macfarlane, 2009; Jacquet & Pauly, 2007; Siemer & Knuth, 2001) but only minimally attend to how such programs politicize, visibilize, and govern food practices. Other strands of research take a more community and place-based approach to understanding how people learn about their relationships with fish, the oceans, and the environment more broadly (Garavito-Bermúdez, 2018; Garavito-Bermúdez & Lundholm, 2017; Patterson, Lindén, Patterson Edward, Wilhelmsson, & Löfgren, 2009; Worster & Abrams, 2005) but also do not explicitly address the politics of these relationships. While this research is important, the current chapter's analysis suggests that environmental education would benefit from more substantive engagement with political ecologies of food, particularly given that the effects of neoliberalism on both education and food systems, and especially their impacts on animal lives, are still not completely understood.

Notes

1. "Crisis-framing" and "innovators" are quoted as these are terms staff utilize, i.e., these are not my terms. For analyses of crisis discourses in EE see Kelsey & Armstrong, (2012) and Russell, Cameron, Socha, & McNinch, (2013).
2. While paid staff who give scripted shows tend to discuss the behavior modifications described above, which are embedded in the scripts, volunteer staff who give unscripted shows tend to favor sharing basic scientific facts about the animals and encouraging visitors to touch animals where allowed. I focus on the role of science in Aquarium teaching practice elsewhere.
3. For brevity, I highlight one exhibit, but other examples of similar neoliberal and disciplinary governance abound at the Aquarium in the data that I collected. I could have also carried out a similar analysis of the Otter Show and the Aquarium's neoliberal and market-based "Adopt a Sea Otter" program.
4. See Kemmerly and Macfarlane (2009) for an analysis of the Monterey Bay Aquarium's (California) "Seafood Watch" program that suggests some positive changes to consumer habits.

References

Agyeman, J., & McEntee, J. (2014). Moving the field of food justice forward through the lens of urban political ecology. *Geography Compass, 8*(3), 211–220.

Allen, P., & Guthman, J. (2006). From "old school" to "farm-to-school": Neoliberalization from the ground up. *Agriculture and Human Values, 23,* 401-415.

Aquarium of the Pacific. (2015). Seafood for the Future overview. Retrieved from http://www.aquariumofpacific.org/seafoodfuture/overview. Accessed May 7, 2018.

Balcombe, J. (2016). *What a fish knows: The inner lives of our underwater cousins.* New York: Scientific American.

Braverman, I. (2013). *Zooland: The institution of captivity.* Stanford, CA: Stanford University Press.

Brown, W. (2003). Neo-liberalism and the end of liberal democracy. *Theory and Event, 7*(1). Retrieved from http://muse.jhu.edu/journals/theory_and_event/v007/7.1brown.html. Accessed September 1, 2020.

Buras, K. L. (2015). *Charter schools, race, and urban spaces: Where the market meets grassroots resistance.* New York, NY: Routledge.

Busa, J. H., & Gardner. R. (2014). Champions of the movement or fair-weather heroes? Individualization and the (a)politics of food. *Antipode 47*(2), 323–341.

Carrier, J. G. (2012). Dollars making sense: Understanding nature in capitalism. *Environment and Society: Advances in Research, 3,* 5–18.

Crist, E. (2013). On the poverty of our nomenclature. *Environmental Humanities, 3,* 129–147.

Crosley, K. L. (2013). Advancing the boundaries of urban environmental education through the food justice movement. *Canadian Journal of Environmental Education, 18,* 46–58.

Drake, L. (2014). Governmentality in urban food production? Following "community" from intentions to outcomes. *Urban Geography, 35*(2), 177–196.

Elgert, L. (2012). Certified discourse? The politics of developing soy certification standards. *Geoforum, 43,* 295–304.

Fletcher, R. (2010). Neoliberal environmentality: Towards a poststructuralist political ecology of the conservation debate. *Conservation and Society, 8*(3), 171–181.

Fletcher, R. (2015). Nature is a nice place to save but I wouldn't want to live there: Environmental education and the ecotourist gaze. *Environmental Education Research, 21*(3), 338–350.

Fletcher, R. (2017). Environmentality unbound: Multiple governmentalities in environmental politics. *Geoforum, 85,* 311–315.

Foster, J. B. (2002). *Ecology against capitalism.* New York NY: Monthly Review Press.

Foster, J. B. (2009). *The ecological revolution.* New York, NY: Monthly Review Press.

Garavito-Bermúdez, D. (2018). Learning ecosystem complexity: A study on small-scale fishers' ecological knowledge generation. *Environmental Education Research, 24*(4), 625–636.

Garavito-Bermúdez, D., & Lundholm, C. (2017). Exploring interconnections between local ecological knowledge, professional identity and sense of place among Swedish fishers. *Environmental Education Research, 23*(5), 627–655/.

Guthman, J. (2007). The Polanyian way? Voluntary food labels as neoliberal governance. *Antipode, 39*(3), 456–478.

Guthman, J. (2008). Neoliberalism and the making of food politics in CA. *Geoforum, 39,* 1171–1183.

Gutierrez, A., & Thornton, F. (2014). Can consumers understand sustainability through seafood eco-labels? A U.S. and U.K. case study. *Sustainability, 6,* 8195–8217.

Hallstein, E., & Villas-Boas, S. B. (2013). Can household consumers save the wild fish? Lessons from a sustainable seafood advisory. *Journal of Environmental Economics Management, 66,* 52–71.

Haluza-DeLay, R. (2013). Educating for environmental justice. In R. B. Stevenson, M. Brody, J. Dillon, & A. E. J. Wals (Eds.), *International handbook of research on environmental* education (pp. 394–403). New York, NY: Routledge.

Hardin, G. (1968). The tragedy of the commons. *Science, 62*(3859), 1243–1248.

Hayes-Conroy, J. (2010). School gardens and "actually existing" neoliberalism. *Humboldt Journal of Social Relations, 33*(1/2), 64–96.

Hobson, K. (2013). On the making of the environmental citizen. *Environmental Politics, 22*(1), 56–72.

Hursch, D. W., Henderson, J. A., & Greenwood, D. (2015). Environmental education in a neoliberal climate. *Environmental Education Research, 21*(3), 299–318.

Jacquet, J. L., & Pauly. D. (2007). The rise of seafood awareness campaigns in an era of collapsing fisheries. *Marine Policy, 31*(3), 308–313.

Jickling, B., & Wals, E. A. J. (2008). Globalization and environmental education: Looking beyond sustainable development. *Journal of Curriculum Studies, 40*(1), 1–21.

Kahn, R. (2011). Towards an animal standpoint: Vegan education and the epistemology of ignorance. In E. Malewksi & N. Jaramillo (Eds.), *Epistemologies of ignorance in education* (pp. 53–70). Charlotte, NC: Information Age.

Kelsey, E., & Armstrong, C. (2012). Finding hope in a world of environmental catastrophe. In A.E.J. Wals & P. Corcoran (Eds.), *Learning for sustainability in times of accelerating change* (pp. 187–200). Wageningin, Netherlands: Wageningin Academic Publishing.

Kemmerly, J. D., & Macfarlane, V. (2009). The elements of a consumer-based initiative in contributing to positive environmental change: Monterey Bay Aquarium's Seafood Watch Program. *Zoo Biology, 28*(5), 398–411.

Larner, W. (2000). Neo-liberalism: Policy, ideology, governmentality. *Studies in Political Economy, 63,* 5–25.

Lemke, T. (2001). "The birth of bio-politics": Michel Foucault's lecture at the Collége de France on neoliberal governmentality. *Economy and Society, 30*(2), 190–207.

Lloro-Bidart, T. (2018). Cultivating affects: A feminist posthumanist analysis of invertebrate and human performativity in an urban community garden. *Emotion, Space, and Society, 27,* 23–30.

Lloro-Bidart, T. (2019). Intersectional & interdisciplinary approaches to interspecies food justice pedagogies. In T. Lloro-Bidart & V. Banschbach (Eds.), *Animals in environmental education: Interdisciplinary approaches to curriculum and pedagogy* (pp. 53–76). New York: Palgrave Macmillan.

Luke, T. W. (1995). On environmentality: Geo-power and eco-knowledge in the discourses of contemporary environmentalism. *Cultural Critique, 31*, 57–81.

McKenzie, M. (2012). Education for y'all: Global neoliberalism and the case for a politics of scale in sustainability education policy. *Policy Futures in Education, 10*(2), 165–177.

McKenzie, M., Bieler, A., & McNeil, R. (2015). Education policy mobility: Reimagining sustainability in neoliberal times. *Environmental Education Research, 21*, 319–337.

Meek, D., & Tarlau, R. (2015). Critical food systems education and the question of race. *Journal of Agriculture, Food Systems, and Community Development, 5*(4), 131–135.

Montgomery, S. (2015). *The soul of an octopus: A surprising exploration into the wonder of consciousness*. New York: Atria.

Mueller, M. J. (2009). Educational reflections on the ecological crisis: Eco-justice, environmentalism, and sustainability. *Science & Education, 18*(8), 1031–1056.

Oakley, J., & Russell, C. (2015). Nourishing ourselves with/in environmental education. *Canadian Journal of Environmental Education, 20*, 5–11.

Oakley, J. (2019). What can an animal liberation perspective contribute to environmental education? In T. Lloro-Bidart & V. Banschbach (Eds.), *Animals in environmental education: Interdisciplinary approaches to curriculum and pedagogy* (pp. 19–34). New York: Palgrave Macmillan.

Patterson, J., Lindén, E., Patterson Edward, J. K., Wilhelmsson, D., & Löfgren, I. (2009). Community-based environmental education in the fishing villages of Tuticorin and its role in conservation of the environment. *Australian Journal of Adult Learning, 48*(2), 382–393.

Potts, A. (2017). What is meat culture? In A. Potts (Ed.). *Meat culture*. Leiden, Netherlands: Brill.

Rice, S. (2013). Three educational problems: The case of eating animals. *Journal of Thought, 48*(2), 112–127.

Rowe, B. (2011). Understanding animals-becoming-meat: Embracing a disturbing education. *Critical Education, 2*(7). Retrieved from https://ices.library.ubc.ca/index.php/criticaled/article/view/182311. Accessed September 1, 2020.

Rowe, B. (2012). Food, habit, and the consumption of animals as educational encounter. *Philosophy of Education*, 210–218. Retrieved from https://educationjournal.web.illinois.edu/archive/index.php/pes/article/view/3627.pdf. Accessed September 1, 2020.

Rowe, B. (2013). It Is about chicken: Chick-fil-A, posthumanist intersectionality, and gastro-aesthetic pedagogy. *Journal of Thought, 48*(2), 89–111.

Rowe, B. (2016). Challenging anthropocentrism in education: Posthumanist intersectionality and eating animals as gastro-aesthetic pedagogy. In S. Rice & A. Rud

(Eds.), *The educational significance of human and non-human animal interactions* (pp. 31–49). New York. NY: Palgrave Macmillan.

Russell, C. (2019). An intersectional approach to teaching and learning about humans and other animals in educational contexts. In T. Lloro-Bidart & V. Banschbach (Eds.), *Animals in environmental education: Interdisciplinary approaches to curriculum and pedagogy* (pp. 35–52). New York: Palgrave Macmillan.

Russell, C., Cameron, E., Socha, T., McNinch, H. (2013). "Fatties cause global warming": Fat pedagogy and environmental education. *Canadian Journal of Environmental Education, 18*, 27–45.

Rutherford, S. (2011). *Governing the wild: Ecotours of power.* Minneapolis: University of Minnesota Press.

Schindel Dimick, A. (2015). Supporting youth to develop environmental citizenship within/against a neoliberal context. *Environmental Education Research, 21*(3), 390–402.

Siemer, W. F., & Knuth, B. K. (2001). Effects of fishing education programs on antecedents of responsible environmental behavior. *Journal of Environmental Education, 32*, 23–29.

Stapleton, S. R., Cole, P., Washburn, M., Jason, M., Alvardo, T. (2017). Views from the classroom: Teacher on food in a low-income urban school district. In I. Werkheiser & Z. Piso (Eds.), *Food justice in US and global contexts* (pp. 129–140). New York: Springer.

Steele, A., & Scott, J. (2016). Emotionality and learning stories: Documenting how we learn what we feel. *Canadian Journal of Environmental Education, 21*(6), 106–124.

Weissman, E. (2015). Entrepreneurial endeavors: (Re)producing neoliberalization through urban youth programming in Brooklyn, New York. *Environmental Education Research, 21*(3), 352–364.

Wells, A. S., Slayton, J., & Scott, J. (2002). Defining democracy in the neoliberal age: Charter school reform and educational consumption. *American Educational Research Journal, 39*(2), 337–361.

Worster, A. M., & Abrams, E. (2005). Sense of place among New England commercial fishermen and organic farmers: Implications for social constructed environmental education. *Environmental Education Research, 11*(5), 522–535.

Chapter Four

Bird Biopower in Lorikeet Forest

Introduction

The Lorikeet Forest, sometimes simply referred to as the Forest, is never quiet. Although the number of people there fluctuates widely, the brightly colored rainbow lorikeets constantly chatter, squawk, and fly about at staggering speeds. Their flight pattern is limited, of course, by an overhanging net extending more than 20 feet in the air. In the 3,200 square foot outdoor habitat, they land on (un)suspecting people, on branches of artificially constructed trees, on the ground where they build nests underneath the dirt, and on the wood fence posts demarcating walking spaces for people. Most of the birds appear especially fond of staff members with whom they have developed a special relationship and visitors who have purchased "liquid gold"—a nectar especially formulated to mimic a natural food source that they slurp up with their evolutionarily-fashioned beaks.[1] The number of birds living in the Forest ebbs and flows as sick, injured, old, and aggressive animals are removed from the exhibit to live temporarily or permanently in "the Barn" behind the Shark Lagoon exhibit and as new chicks hatch and are released.

Besides visitors who travel in and out of the Forest each day, usually for only a few minutes of time, the lorikeets also have regular contact with Aquarium staff members, some of whom become recognizable to the birds. The first few months I ventured into the Forest, like any novice, I would jerk slightly to the right or left as I felt a lorikeet come rushing toward what I thought was my head. Trained staff explained that when human interlocutors move suddenly, the likelihood of a collision occurring increases. The lorikeets more deftly avoid crashes with non-moving objects than moving ones, making my sudden movements possibly dangerous for us both. Aquarium staff work tirelessly to prevent these interactions by carefully monitoring, regulating, and surveying both the animals who live at the Aquarium and the over one million visitors who come to the facility each year. In this chapter I show how the Aquarium, its staff, and its animals co-participate in a complex neoliberal edutainment project wherein the institution governs lorikeet bodies through multiple disciplinary and regulatory technologies crafted to ensure visitors have a satisfying experience and become more conservation-minded. These disciplinary practices include: (1) staff quietly surveying and/or verbally admonishing visitors to regulate behavior; (2) lorikeets biting, defecating on, and refusing to interact with visitors; and (3) the institution requiring that lorikeets be fed a modified diet during the day and that they remain captive. The lorikeets are also managed biopolitically as staff monitor the health and numbers of the population, or "make live" through disease management and breeding practices. Yet, since this "making live" is also linked to the fiscal survivability of the institution, these practices reflect a neoliberal exercise of biopower (Fletcher, 2010). I also borrow from Bruno Latour (1987), Michel Callon (1986), and Donna Haraway (1989, 2008) implicating not only humans, but nonhuman animals as actors.

The management of lorikeet bodies, for example, is linked to the fiscal survivability of the institution that Rob Fletcher (2010) describes as a neoliberal exercise of biopower since the emphasis on economics appears to be equally as important as that of sustaining life, and perhaps more so. Further, the Aquarium's conservation education project imagines visitors as individualized consumers who have the freedom to choose their edutainment experiences (e.g., coming to the Aquarium to tactually interact with lorikeets or small sharks), as well as their environmentally responsible behaviors, as discussed in Chapter Three.

Yet the neoliberal aspirations of the institution and many of its staff are not enacted without a hitch. As I demonstrate in this chapter and in Chapter Six, lorikeets, as well as other Aquarium birds, are agentic beings who complicate these encounters in manifold ways, which include biting, refusing to sip nectar and/or tactually interact, and defecating. My use of the term "agency" here is not meant to imply that lorikeet intentionality is the same as the exercise of human agency. In Helena Pedersen's analysis of a research scientist, Emma and the battery hens she studied, for example, it was clear that the hens could not escape their "forced labour," but they were able to "complicate, disturb, delay, or change the order of the working scheme" (2011, p. 19) as they expressed limited forms of agency. I therefore utilize "agency" through an animal performativity framework (Hovorka, 2015) to demonstrate that nonhumans and humans co-produce agencies and relationships as they respond to each other's actions, albeit utilizing different bodies and means of communication.

I begin this chapter with a brief description of the exhibit background and lorikeet natural history to orient the reader to lorikeet experiences both at the Aquarium and in the wild. Next, I discuss how lorikeet nectar is used as a biopolitical tool to regulate lorikeet bodies and provide visitors with the kinds of human-animal interactions they have come to expect at institutions like the Aquarium. I then unpack how the Aquarium also regulates visitor action not only to ensure lorikeet and visitor safety, but also to provide a space where visitors can have a good time. Before concluding, I discuss how human-lorikeet interactions in the Forest are heavily shaped by the lorikeets themselves as they mediate interaction and sometimes defecate on or bite visitors. And finally, I assert that commodified tactual encounters between visitors and lorikeets are unlikely to foster more flourishing or ethical relationships with other animals, which I explore more fully in Chapters Six and Seven.

Exhibit Background

In 2003, five years after the Aquarium opened, the Forest was constructed. Originally a lush, green, and forested aviary, at the time of my fieldwork the habitat appeared denuded, with constructed "trees" pieced together from dried out branches anchored to the floor and covered with ground cover, pebbles, and grass to appear natural. From these "trees" hang bird toys resembling those one would see in a pet store. According to an Aquarium staff member, birds destroyed the live trees originally present in the Forest so the trees were removed. The Forest floor consists of a concrete pathway where visitors are permitted to walk, flanked on both sides by areas forbidden to visitors. Lorikeets generally build their nests in these spaces, either on the ground or in the provided nesting boxes. A large overhanging net extending 20 feet into the air prevents the lorikeets from escaping (see Figure 4.1).[2]

Figure 4.1: Constructed trees inside the Lorikeet Forest.

Lorikeet Natural History

Rainbow lorikeets are native to Australia and the South Pacific and are considered pests in Western Australia (Massam & Mawson, 2009).[3] At the time of my fieldwork, the exhibit featured five of the 22 subspecies of rainbow lorikeets that exist in the wild. According to the International Union for Conservation of Nature (IUCN), rainbow lorikeets (*Trichoglossus haematodus*) are a species of Least Concern due to their large range and population size, despite declining populations in some areas. Since 1981, when they were listed on CITES Appendix II (indicating that they are not threatened or endangered, but should be monitored to ensure that wildlife trade does not threaten their long-term survival), over 100,000 individuals have been caught up in international trade.[4] While it is not known whether these animals ended up in aquariums, zoos, or as exotic pets, these uses all commodify wildlife and the "animals' lives are a shadow of their previous existence" (Collard & Dempsey, 2013, p. 2686; see also Collard, 2014; Nxumalo & Pacini-Ketchbaw, 2017).

Figure 4.2: Rainbow lorikeet *(Trichoglossus haematodus)*.

Biopolitics: Biological and Spatial Management as "Edutainment"

Staff overwhelmingly believe visitors come to the Aquarium to be entertained, primarily through interactions with animals and staff. These assumptions are grounded in evidence from annual Visitor Survey Reports, where visitors during the summer of 2013 cited entertainment (94 percent) and interest in marine life (86 percent) as their top reasons for visiting, followed by education (52 percent).[5] Education, therefore, is often "slipped in" and related to the entertainment experience, as a staff member relayed to me in an interview:

> *I asked Charlotte why she thinks people visit the Aquarium:*
>
> *I think, for example, our penguin exhibit opened in May. Just certain animals that people really like, you know,"HUH! They have penguins?" So I think people want to see the different animals, the different types of fish. I think in a sense they want to be entertained. I think that's where the teaching comes in. You have to entertain them and inform them at the same time. For me, I say, learning to be fun. I'm teaching them, but I want them to have fun while they're learning. Certain animals draw people. Shark Lagoon is popular. The lorikeets are very popular.*

As illustrated above, the institution and its staff postulate that visitors, envisaged as neoliberal consumers of edutainment experiences "seeking to fulfill themselves with a variety of micro-moral domains or 'communities' " (Rose, 1996, p. 57), come to the Aquarium to be entertained, in many cases by animals. As such, visitors are thought to engage in ongoing cost-benefit analyses as they decide where their leisure and education dollars will be disbursed (Fletcher, 2010; Lemke, 2001). This knowledge drives many decisions and concretely affects the lives of the lorikeets as the institution and its staff manufacture the kinds of human-animal interactions they believe visitors desire.

When visitors approach the Forest, two Guest Services volunteer staff members greet them. One monitors the entrance and the other stands behind what looks like a tiki bar at a tropical island getaway, presumably to add to the theme park atmosphere. Rather than advertising nightly drink specials, however, this one displays a sign that reads, "Feed nectar to the birds" or "Alimente los periquitos con nectar,"

"$3."[6] Another sign to the left, entitled "Lorikeet Hunger Meter," indicates to visitors how hungry the birds are. A different one to the right lists the over 50 ingredients found in the nectar, specially prepared by Husbandry staff to "mimic nature." (See Figure 4.3).

Figure 4.3: Lorikeet nectar bar.

Besides the ticket (or membership) that visitors must purchase to gain entry into the Aquarium, the purchase of nectar represents the first commodifiable and regulated interaction they can have with the lorikeets. By selling the nectar to visitors, the Aquarium gains financially and, in effect, controls human-lorikeet interactions—only those visitors who choose to purchase it typically have body-to-body contact with the birds. The lorikeets do not usually land on humans unless they have food, and visitors are not allowed to grab, touch, or otherwise force physical interaction, partly because most lorikeets would bite them and inflict serious physical harm if they did so. As Jane Desmond explains in her book, *Staging Tourism*, animal reliance on people for food in zoos and aquariums is typical despite the attempt to create a space that seems "natural." She comments,

> The illusion is that we are seeing not only authentic animals but authentic performances of species-species behavior as well. In fact, what we do see in this display of wildness is a display of total dependence on humans for food, care, protection, and survival. (1999, p. 164)

Unlike most zoo and aquarium exhibits, the Forest lorikeets become totally dependent not on their usual staff caretakers, but on visitors seeking a short-lived experience with wild animals.[7] In this instance, however, the nectar is more than just the means through which contact occurs. Here, it becomes a biopolitical tool through which neoliberal governmentality acts. Nectar is not so much used to provide the birds' sustenance (which would be a disciplinary approach to biopolitical management) since Husbandry staff members feed off-exhibit lorikeets a heavier "smoothie" during the day and on-exhibit lorikeets the same smoothie at night. Instead, nectar is a carefully crafted neoliberal biopolitical tool designed to support economic growth—in this case, that of the institution. This biopolitical tool "works" primarily through two means: (1) it dilutes the lorikeet's feces so that they are less offensive to visitors; and (2) it fosters tactual encounters with lorikeets who would otherwise not interact with humans.

To ensure visitor satisfaction, exhibit lorikeets are fed nectar during the day to make their feces water-like and less viscous so that when they defecate on unlucky visitors, it makes it less likely their momentary animal companions will offend them. Here is an excerpt from my field notes from an encounter with a staff member, Jennie, while she worked inside the Forest:

> *At this point the Husbandry staff member and I are standing right at the edge of the entrance to the Hut, where there are ropes to signal to visitors that this is a forbidden area. The Hut is small structure near the exit of the Forest where staff temporarily house, in small, wire mesh bird cages, a small number of injured birds or juvenile birds who will eventually be released into the Forest or permanently taken off exhibit to live in the Barn. I was afraid of getting pooped on by one of the birds perched on the roof of the Hut because I saw it happen to a visitor a few minutes earlier.*

Teresa: I was looking up because one of them projectile pooped on a little girl (laughing). It was like water. Oh, it peed on you. That's what they said. Because it was like water.

Jennie: What happens is that we don't feed them the smoothie in here. The ones in the Barn (where Lorikeets are housed off-exhibit) get it during the day. But they specifically don't give it to these guys until after dark, or 'til closing, so the poop won't get more obnoxious for people.

Teresa: Oh I see.

Jennie: So it's that liquid, loosey stuff. And then at night they get the smoothie, their more substantial meal. It makes their poop a little more—

Teresa: Thicker?

Jennie: Yeah. So they have to feed them differently in here, just so—

Teresa: To accommodate the guests.

In addition, the nectar is a lighter meal that does not fully satiate the birds, which enables a greater number of visitors to purchase nectar and have interactions with the lorikeets the institution imagines they crave. Further, it increases the amount of nectar the institution can sell each day, which contributes to fiscal solvency. The Aquarium, by specially designing this meal to make the birds' feces less offensive and then selling the nectar to visitors, thus biopolitically regulates the lorikeet population on display in the Forest, controlling and commodifying both the birds and their interactions with humans.

In an analysis of the shift of zoos (and aquariums) from institutions that promoted the visitor "gaze" to institutions dedicated to education and entertainment, Alan Beardsworth and Alan Bryman (2001) point out that wild animals in zoos and aquariums are "sanitized" in what they call the process of commoditization (see also Higham & Neves, 2015; Fletcher, 2014; Russell & Ankenman, 1996; Rutherford, 2011; Wearing & Jobberns, 2011). Through spatial and bodily regulation the Aquarium packages the birds as an *interactive*, and not just visual, product that can be sold to visitors for financial gain. Although the habitat and their food have been constructed to mimic nature, elements of true wildness are missing there. Unable to forage, the birds become dependent on humans for food; they cannot fly about and seek a mate elsewhere, so they must choose a mate from the other birds in the Forest; and they have become, over time, habituated to the presence of human beings.

"STOP THAT! DON'T DO THAT!" and "Keeping an Eye on the Kids"

Forest staff discipline visitor behavior in multifaceted ways to ensure they have a positive experience and that the lorikeets stay "safe," which is typically defined as not escaping as well as not being grabbed or poked. While stationed inside the Forest, staff simultaneously bustle about the space and gaze over visitors as if they were security or police officers surveying their actions. Some staff members actually describe their roles as "babysitters" or "policemen." When necessary, they respond to inappropriate actions, which might include polite or scathing verbal admonitions followed by explanations as to why the behavior is inappropriate and, very rarely, by using their own body as a means to grab the attention of the offending visitor. In order to ensure a satisfying experience for visitors, staff are trained that the primary means they should interact with visitors is through politely and positively suggesting correct behaviors rather than scolding them for improper actions. While Education staff on the microphone may appear to be giving visitors a "show" or Husbandry staff feeding the birds in the Hut may seem as though they are not paying attention to visitors as they bustle about, they are constantly surveying visitor behavior.

The field note excerpt below illustrates how the physical space necessitates certain staff actions given that there are areas where birds are permitted, but visitors are prohibited:

> *I noticed some of the kids tried to climb over or under the short wood fence separating the concrete area from the grassy play area for the birds. The Education volunteers (there were several present at that time), Husbandry volunteer, and the paid Education staff member all ran around like busy bees trying to keep the kids from entering these forbidden areas. In all interactions I observed, the staff members were polite when asking visitors to obey certain rules, i.e., they used "please," did not yell, and explained themselves. Several times I also noted parents reinforcing the rules after hearing the Aquarium staff member explain why they exist—animal safety and human safety. Jennie, with her deep commitment and love for the birds, was probably the busiest as every time she noticed someone potentially misbehaving (e.g., crawling or climbing where they shouldn't or trying to poke the birds with a finger), she rushed over and gently told the visitors how to interact properly with the animals.*

During my time at the Aquarium I did not observe any staff overtly scolding visitors, although several told stories of when they felt compelled to utilize more forceful means. Institutionally, forceful actions are undesirable because they are believed to negatively affect the visitor experience and could, therefore, fiscally harm the Aquarium. In this way, staff, like lorikeets, are carefully managed in order to cater to visitors, who sustain the institution through ticket sales and other purchases.

The following excerpt depicts a conversation I had with Sharon and Jennie in the Forest one morning. Since both have worked in the Education and Husbandry departments, I asked them to describe how those roles might be different since several staff members had told me that each department has different unwritten rules:

> *Sharon: We're [Husbandry] expected to be diplomatic, but if it comes between an animal being injured and a guest having a good experience, we pick the animal.*
>
> *Jennie: And the Education people are a little more reluctant to do that because they …*
>
> *Sharon: They get in trouble if they do. That's the thing. They're not supposed to.*
>
> *Jennie: We're supposed to be diplomatic, too. And I try just as hard to be diplomatic in Husbandry as I do in Education, but to be honest with you I will honestly, hands down, protect the animal and try not to, I try to make the guest, nothing negative with the guest, but if it comes to I having to choose—oh well. I'm not going to let the animal get hurt.*

Note that in the above passage, both Sharon and Jennie acknowledge the importance of "being diplomatic"; in another excerpt Sharon discusses avoiding "customer complaints." These examples illustrate the tension at the Aquarium between privileging the experience of visitors versus that of the lorikeets. In this space, the typical visitor gaze upon confined wild animals (Berger, 2009; Desmond, 1999; Milstein, 2009) has metamorphosed into a different kind of asymmetrical human-animal relation (Haraway, 2008) that necessitates third-party regulation precisely because the Aquarium uses nectar as a biopolitical tool to foster tactual interaction with confined wild animals. Visitors are imagined as craving a short-lived experience with the "wild," which the institution heavily promotes, but they must do so in a prescribed fashion to ensure the safety of the lorikeets and themselves.

The institutional focus on visitor satisfaction is at root an anthropocentric educational endeavor (Lloro-Bidart, 2015; Rowe, 2016; Russell & Bell, 1996) as it constructs an experience designed to meet human needs and wants. Through bioregulating and commodifying animal bodies and nature (Rutherford, 2011), the institution hopes to guarantee its own existence. The staff who work directly with the birds in the Forest, however, embrace a different approach and view themselves as "protectors" of the animals, as I will describe in more detail later in this chapter and again in Chapter Six.

Lorikeets Discipline Their Human Companions

The prior section described how the Aquarium offers visitors access to some of the wild aspects of nature they are presumed to desire, but in an anthropomorphized and restrained way as bird bodies are biopolitically regulated to fiscally benefit the institution. Yet lorikeets, as wild living beings, disrupt human aims in many ways. They bite staff and visitors, they refuse to tactually interact (sometimes even with those visitors who have purchased nectar), and their bodies must defecate despite the institution's attempt to regulate this defecation through only providing nectar as sustenance throughout the day. These forms of lorikeet "resistance," as demonstrated below, sometimes complicate visitor, and even staff, satisfaction.

Be Careful! They Might Bite!

On a busy day, one can spy rows and rows of strollers parked in front of the doors of the Forest, many overflowing with all of the typical accouterments of American toddlers and their parents—spare diapers and items of clothing, toys, blankets, baby wipes, and rattles. According to a volunteer educator, some years ago management staff decided strollers pose a hazard inside the habitat since lorikeets can unknowingly crawl into the carriage of the strollers. Apparently, trapped lorikeets have bitten unlucky staff members tasked with removing them so hard they drew blood. As a result, the staff member at the entrance of the Forest informs visitors that if they refuse to comply, they cannot enter the space. Lorikeet biting, therefore, serves as a mechanism to regulate human behavior. People wishing to have physical contact with the birds or to simply venture into their home must abandon their strollers outside the habitat or be denied the opportunity to interact.

The threat of a lorikeet bite also deters some staff members from taking shifts in the Forest. In an interview, Rose, a staff member, describes a lorikeet "savagely" and repeatedly biting her as she accidentally stepped on its nest one day. Interestingly she did not blame the lorikeet, but instead acknowledged that the unnatural human-lorikeet contact in the Forest was to blame. In the excerpt below, notice how she cites an inability to mistreat the bird, only partially because visitors are watching, despite being in physical pain:

Teresa: Is there any exhibit that you won't go to?

Rose: There is one that is not my favorite and my day captain, Vera, she knows I will [only] go there if I have to. We are sitting right next to it.

Teresa: The birds?

Rose: The lorikeets.

I laugh because everyone seems to have a strong opinion about the birds.

Rose: I was attacked. And it was my fault. This is called getting used to the fact that these are wild animals and they do have their patterns of behavior and getting to understand them. And I was in there and I stepped in front of a brood box that had a pair inside, a breeding pair, and the male didn't like that and he came out and attached himself to my ankle. And it drew blood. And you have to be really professional. I know I'm free [she's a volunteer staff member], but I still have to be professional.

Teresa: Yeah.

Rose: So I couldn't kick him off because you can't mistreat him. You can't swear at him, you can't yell at him. People are going, "You have a bird on your …" "I know I have a bird. And I know I'm bleeding." So I finally went like that [shows how she shooed him away with her hand] and got him off, but then I got him on my hand and so he drew blood there, too. So, that's why I don't [like to go there], but I do, I take my people in there when we have the tours. I go through and tell them my story. And I tell them, "I hope you like it. Because a lot of people love it." And they do.

Teresa: Yeah. The people I've talked to either love, love, love the birds or "No thank you."

Rose: I was okay up until that and then I was like, "No thank you."

She goes on to explain that she still loves the birds and even shows me a piece of "bird" jewelry she likes to wear. However, due to her bloody interactions, Rose no longer takes shifts in the Forest if she can avoid them.

Rose's case illustrates that the packaging of human-lorikeet interactivity requires a kind of biopolitical regulation that renders lorikeet bodies consumable; as vital, living beings however, they are not entirely subject to manipulation. In contrast to Luke's (2000) historical analysis of the Audubon Society, which, through what he called "technologies of Audubonization," constructed an innocuous, anthropomorphized, and romantic vision of wild birds with the hope that people would care about and want to protect them, the Aquarium constructs a beautiful lorikeet, but certainly not a defenseless one. The continual warning that birds may bite, and the sharing of stories like that of Rose's above illustrate that they have bitten people in the past. Whereas the Audubon Society sought to modify human-bird interactions through advocating gazing upon and photographing aesthetically pleasing and gentle creatures (Luke, 2000), the Aquarium actually promotes body-to-body contact through what I have termed technologies of touch. The institution, therefore, sanitizes biological and spatial aspects of wildness that are amenable to regulation, while simultaneously warning humans of behaviors staff cannot totally regulate (i.e., biting). These technologies of touch aim to provide visitors with enjoyment at the Aquarium so that they will have a positive experience.

"They Just Won't Land on Me!" "Get These Birds OFF of Me!"

At any given time, birds continually fly about and land on trees as they feed, groom, nest, and occasionally, depending on the season, mate. As visitors make their way through the habitat, some stop to take photos with the lorikeets or film them as they carry out their daily activities, others stop to read the placards flanking each side, and those who purchase food attempt (usually successfully) to feed them. While some visitors are bombarded with lorikeets as soon as they enter with food, others must be more patient in order to have their desired interaction.

The prohibition of touching lessens the likelihood that the lorikeets will bite visitors and ensures that visitors do not improperly handle the lorikeets. Visitors may attempt to feed the lorikeets; however, the decision to land on a human and sip up nectar is entirely up to the individual bird. If the lorikeets are not hungry or seem disinterested in food, the visitor must seek out more famished birds and wait for physical contact. The lorikeets, in this way, participate actively in creating the kinds of interactions visitors can have with them. That is, the actual interaction is in this way lorikeet-driven despite the Aquarium's attempt to spatially regulate and manage the lorikeets to make them more accessible and acceptable to visitors who have paid for certain kinds of interactions. This excerpt from my field notes is illustrative:

> *People carried around their little vessels of nectar in ice-cream cone fashion, patiently waiting for the birds to approach them and lick up the delicious treat with their tongues that evolved over time to be adept at lapping up nectar from plants. Of course, like I have seen every day, there were folks who waited patiently for birds to come up to them, but their wait was to no avail. The birds, for whatever reason (perhaps lack of hunger or perhaps lorikeet intuition about which humans are "safe" and which are not) would not come near, while other folks were nearly toppled over by four or five birds that landed on any available extremity, including the head. At one point today, there was a man so covered in birds that others gathered around him to watch, some taking photos and others filming. It was quite the spectacle.*

Staff members support this avian-initiated disciplining, as my conversation with a staff member in the Forest illustrates:

> *There are different ways to interact with them. And what we tell the guests, I tell the guests, is they [the lorikeets] consider fingers aggression. Don't use your fingers. This is an invitation (she holds her arm out the way she would if she wanted a bird to get on her arm to sit). So we teach them how to interact. And then the bird gets to say, "Yeah. I'd like to come see you" or "No. I don't want to." And this is their house and they get to make the rules.*

Conversations with visitors indicate that many are aware that the birds mediate their interactions. In this interview, I chat with Jose and Leticia, a young married couple with children, inquiring about the exhibits they visited:

Teresa: Oh, you went in the Lorikeet Forest?

José: Yeah. And we got to feed them too, with the special liquid. I don't know what it was.

Teresa: Did you guys get to touch them at all?

José: Actually, when you touch them, they kind of go to the side and they kind of want to fly away. But I managed to kind of touch one while it was eating.

Teresa: So you had the food? Did you have the food in your hand and they ended up on your arm?

José: Yeah. You have to grab it like an ice cream cone [he shows me with his hands and arm]. And you have a little, yeah, and they just come to you. They just fly and go to you.

Leticia: It was funny 'cause, like, I wouldn't, like, go to them. I don't know. They would, like, push me and they would land on top of me.

José: Yeah, see. They were flying to her. And I was right there alone. And then I got like four of them coming to me. So I had a bunch of birds on me.

The above passage illustrates that most visitors, as they traverse through the Forest, have understood how to feed the birds, realize the interaction is lorikeet-dependent, and are willing to be patient. Most visitors do not, however, interact with a staff member who has knowledge about the birds; there is no forum for visitors to question why the birds are held captive; visitors are not at all involved in animal rehabilitation; and visitors are not aware that their feeding the birds often disrupts lorikeet territorial behaviors.

"Good Luck Polka Dots"

In *The History of Sexuality*, Michel Foucault (1979) discussed the state's regulation of bodies in the name of health and hygiene. In modern society, hygiene has come to mean maintaining the body as clean as possible, and therefore free of potential disease-causing agents, many of which are associated, rightly or wrongly, with feces (Curtis & Biran, 2001). Many visitors and staff, like most people, have therefore developed a cultural aversion to feces. The threat of being defecated on is significant enough that some volunteer educators refuse to take a shift inside the Forest, despite the fact that the paid staff members who supervise volunteer training ask volunteer staff members-in-the-making to convince visitors that lorikeet poop is innocuous because, "It's just nectar. So they shouldn't feel bad because it would be worse if they pooped out fish." This "threat," of course, is nothing new when one considers the theme of avoiding bird poop occurs in countless films, novels, and cartoon strips, and recently became a concern due to avian flu epidemics caused, in part, by humans who had contact with infected bird feces. An online search of "bird poop" reveals myriad sites dedicated to selling cleaning products specially designed to remove the feces and halt damage to clothing or paint (on cars, for example).

Kristal, a volunteer staff member who refuses to work in the Forest explains that the bird feces damage clothes:

Teresa: No birds for you either?

Kristal: No.

Teresa: Did you have a bad experience with the birds or you're just not into the birds?

Kristal: I'm not into poop. These guys projectile poop.

Teresa: Yeah.

Kristal: At least 12 feet. It's got a bleach-like quality. So, most of the people that work in there have these little white spots on there, bleach-outed. "Good luck polka dots," they call them.

While staff members like Kristal refuse to work in the Forest, some staff and visitors are willing to take their chances. Below, Sydney and Cisco, both first time visitors, discuss how they do not mind the feces; to them it's what one expects in an interaction with an animal. They express, however, being slightly uncomfortable with how fast the body-to-body and feces-to-body contact took place once they were inside. As illustrated in the interview excerpt below, lorikeets, as captive wild animals, can exercise agency and moderate human behavior, vis-à-vis these interactions:

Teresa: So could you tell me about what happened when you guys walked in [to the Forest] and you had the food in your hands, like, tell me about your experience in there?

Cisco: I got attacked as soon as I walked in. A bird already got on me and did his duty on me and then he took off after that (laughing).

Teresa: Oh. Did he poop on you?

Cisco: (laughing) Yeah.

Sydney laughs, too.

Teresa: So did you know going in they were probably going to land on you since you bought food?

Cisco: Yeah.

Teresa: So you were comfortable?

Sydney: Not that fast, though.

Teresa: Not that fast?

Cisco: I knew it could've happened.

Teresa: And you were okay with being pooped on?

Cisco: Yeah. It's just a bird. That's what he does.

Other visitors, like Paula, Stacey, and Bobby, treaded lightly into the Forest. Stacey, a college student, and Bobby, a senior in high school, visited the Aquarium with a family friend they called Aunt Paula. It was their first visit, but Paula had been several times prior. For them, refusing to purchase nectar decreased the likelihood of being defecated on, so they took photos as they passed through, but did not purchase nectar.

As the staff members and Aquarium visitors discussed above demonstrate, no universal "modern attitude" toward bird feces exists. Attitudes toward "matter out of place" (i.e., dirt, excrement, menstrual blood) have been variously shaped by constructed categories that are "public matters" (Douglas, 1966, pp. 36–41). In *The Civilizing Process*, Norbert Elias (1978) traces the history of development of aversions to "natural functions" that increasingly became isolated from public view. This concealment occurred through complicated processes of social change whereby it became largely inappropriate to defecate, urinate, flatulate, or belch in public in Western societies. Therefore, some Aquarium visitors, although seeking contact with "wildness," expect a sanitized version that will not upset their notions of contemporary "dirt" avoidance, systemically tied to hygiene, aesthetics, knowledge of pathogenic organisms, and more importantly, for Mary Douglas, notions of "matter out of place" (1966, p. 36). However, as some of the interview excerpts above illustrate, particularly that of Sydney and Cisco, bird feces do not put off all Aquarium visitors. The Aquarium as an institution assumes, though, that most visitors would be offended and attempts to regulate bird bodies as a result.

Conclusion

Zoos and aquariums have endured sustained criticism because they have historically encouraged visitors to "gaze" over captive wild animals in simulated habitats (Beardsworth & Bryman, 2001; Milstein, 2009), encouraging the Cartesian separation of humans and nature (Milton, 2002, 2005; Williams, 1980). The Aquarium has sought to address this criticism by imagining a visitor craving to put the "human" back in "nature" through lorikeet and human body-to-body contact. Seemingly, the goal is to deliver this contact, while simultaneously fostering more ethical human-animal relationships and ensuring the institution's prosperity. However, I argue that body-to-body contact cannot substitute for sustained contact with wild animals or nature, more generally, or contact in a space less commodified and regulated than that of the Lorikeet Forest. In Chapter Five, I will demonstrate how small shark bodies are similarly biopolitically regulated. Later, in Chapters Six and Seven, I will explore how staff, whose experiences with the lorikeets and other Aquarium birds are qualitatively different than that of visitors, come to develop intersubjective relationships with the birds. Further, I consider what it might mean for zoos and aquariums to actually resist the commodification of wildlife.

Notes

1. In the wild, lorikeets eat nectar, blossoms, buds, and fruit. According to signage at the exhibit, the nectar sold to visitors is designed to mimic what they would get in the wild. That same sign lists the 30+ ingredients in the nectar.
2. See Karen Morin (2015) for a discussion of the similarities between prisons and zoo and aquarium cages.

3. http://www.iucnredlist.org/details/160032124/0
4. The IUCN defines a species of Least Concern in the following manner: "A taxon is Least Concern when it has been evaluated against the criteria and does not qualify for Critically Endangered, Endangered, Vulnerable or Near Threatened. Widespread and abundant taxa are included in this category." (http://www.iucnredlist.org/static/categories_criteria_3_1). Other lorikeet species are variously listed as Critically Endangered, Endangered, Threatened, Vulnerable, or of Least Concern.
5. In addition to the statistics cited in the text, in the summer of 2013, visitors also cited the following reasons for visiting the Aquarium: sightseeing (70 percent); curious to see (62 percent); bring children (57 percent); other (18 percent); because I'm a member (16 percent). (Aquarium of the Pacific Summer Survey Report, 2013).
6. Lorikeet bodies also generate revenue for the institution through nectar sales. In 2011, nectar sales earned $389,454 for the Aquarium, which constitutes 2% of Program Services revenue and 1.4% of total revenue (Aquarium of the Pacific 990 Tax Form, 2011).
7. John Knight's analysis of monkey parks in Japan revealed a similar phenomenon. Visitors there enter into a food exchange with monkeys, producing human-animal interactions that would be impossible without food (Knight, 2005).

References

Beardsworth, A., & Bryman, A. (2001). The wild animal in late modernity: The case of the Disneyization of zoos. *Tourist Studies*, *1*(1), 83–104.

Berger, J. (2009). *Why look at animals?* New York, NY: Penguin Books.

Blue, G., & Alexander, S. (2015). Coyotes in the city: Gastro-ethical encounters in a more-than-human world. In K. Gillespie & R. Collard (Eds.), *Critical animal geographies: Politics, intersections, and hierarchies in a multispecies world* (pp. 194–212). New York, NY: Routledge.

Callon, M. (1986). Some elements of a sociology of translation: Domestication of the scallops and the fisherman of St. Brieuc Bay. In J. Law (Ed.), *Power, action, and belief: A new sociology of knowledge?* (pp. 196–223). London, UK: Routledge.

Collard, R-C. (2014). Putting animals back together, taking commodities apart. *Annals of the Association of American Geographers*, *10*(4), 151–165.

Collard, R-C., & Dempsey, J. (2013). Life for sale? The politics of lively commodities. *Environment and Planning A*, *45*, 2682–2699.

Curtis, V., & Biran, A. (2001). Dirt, disgust, and disease: Is hygiene in our genes? *Perspectives in Biology and Medicine*, *44*(1), 17–31.

Desmond, J. C. (1999). *Staging tourism: Bodies on display from Waikiki to Sea World.* Chicago, IL: University of Chicago Press.

Devall, B., & Sessions, G. (1985). *Deep ecology: Living as if nature mattered.* Salt Lake City, UT: Peregrine Smith Books.

Douglas, M. (1966). *Purity and danger.* London: Routledge and Kegan Paul. *Geoforum, 43,* 295–304.

Elias, N. (1994). *The civilizing process.* Oxford: Basil Blackwell Ltd.

Fletcher, R. (2010). Neoliberal environmentality: Towards a postructuralist political ecology of the conservation debate. *Conservation and Society, 8*(3), 171–181.

Fletcher, R. (2014). *Romancing the wild: Cultural dimensions of ecotourism.* Durham, NC: Duke University Press.

Haraway, D. (2008). *When species meet.* Minneapolis, MN: University of Minnesota Press.

Haraway, D. J. (1989). *Primate visions: Gender, race, and nature in modern science.* New York, NY: Routledge.

Hovorka, A. (2015). The *gender, place, and culture* Jan Monk distinguished annual lecture: Feminism and animals: Exploring interspecies relations through intersectionality, performativity, and standpoint. *Gender, Place, & Culture, 22*(1), 1–19.

Knight, J. (2005). Feeding Mr Monkey: Cross-species food 'exchange' in Japanese monkey parks. In J. Knight (Ed.), *Animals in person: Cultural perspectives on human-animal intimacies* (pp. 231–253). New York, NY: Berg.

Latour, B. (1987). *Science in action: How to follow scientists and engineers through society.* Cambridge, UK: Harvard University Press.

Lemke, T. (2001). "The birth of bio-politics": Michel Foucault's lecture at the Collége de France on neoliberal governmentality. *Economy and Society, 30*(2), 190–207.

Lloro-Bidart, T. (2015). A political ecology of education in/for the Anthropocene. *Environment and Society: Advances in Research, 6,* 128–148.

Lloro-Bidart, T. & Russell, C. (2017). Learning science in aquariums and on whale-watching boats: The political deployment of other animals. In M. P. Mueller, D. J. Tippins, & A. J. Stewart (Eds.), *Animals in science education: Ethics, curriculum, and pedagogy* (pp. 41–50). New York, NY: Springer.

Luke, T. W. (2000). Beyond birds: Biopower and birdwatching in the world of Audubon. *Capitalism Nature Socialism, 11*(3), 7–37.

Massam, M., & Mawson, P. (2009). Battling bird damage. *Australian Viticulture, 13*(6), 16–23.

Milstein, T. (2009). "Somethin' tells me it's all happening at the zoo": Discourse, power, and conservationism. *Environmental Communication, 3*(1), 25–48.

Milton, K. (2002). *Loving nature: Towards an ecology of emotion.* New York, NY: Routledge.

Milton, K. (2005). Anthropomorphism or egomorphism? The perception of non-human persons by human ones. In J. Knight (Ed.), *Animals in person: Cultural perspectives on human-animal intimacies* (pp. 255–269). New York, NY: Berg.

Morin, K. M. (2015). Wildspace: The cage, the supermax, and the zoo. In K. Gillespie & R. Collard (Eds.), *Critical animal geographies: Politics, intersections, and hierarchies in a multispecies world* (pp. 101–125). New York, NY: Routledge.

Nxumalo, F., & Pacini-Ketchabaw, V. (2017). "Staying with the trouble" in child-insect-educator common worlds. *Environmental Education Research, 23*(10), 1414–1426.

Pedersen, H. (2011). Counting affects: Mo(ve)ments of intensity in critical avian education. *Canadian Journal of Environmental Education, 16*, 14–28.

Rose, N. (1996). Governing "advanced" liberal democracies. In A. Barry, T. Osborne, & N. Rose (Eds.), *Foucault and political reason* (pp. 37–64). Chicago, IL: The University of Chicago Press.

Rowe, B. (2016). Challenging anthropocentrism in education: Posthumanist intersectionality and eating animals as gastro-aesthetic pedagogy. In S. Rice & A. Rud (Eds.), *The educational significance of human and non-human animal interactions* (pp. 31–49). New York, NY: Palgrave Macmillan.

Russell, C. L., & Ankenman, M. J. (1996). Orangutans as photographic collectibles. Ecotourism and the commodification of nature. *Tourism Recreation Research, 21*(1), 71–78.

Russell, C. L., & Bell, A. C. (1996). A politicized ethic of care: Environmental education from an ecofeminist perspective. In K. Warren (Ed.), *Women's voices in experiential education* (pp. 172–181). Dubuque, IA: Kendall Hunt.

Rutherford, S. (2011). *Governing the wild: Ecotours of power*. Minneapolis, MN: University of Minnesota Press.

Wearing, S., & Jobberns, C. (2011). Ecotourism and the commodification of wildlife. In W. Frost (Ed.), *Zoos and tourism: Conservation, education, entertainment?* (pp. 47–58). London, UK: Channel View Publications.

Williams, R. (1980). *Problems in materialism and culture*. London, UK: Verso.

Chapter Five

The Disembodied Shark

Introduction

As Aquarium visitors approach the Shark Lagoon, the image of the gigantic, human-eating predator caricatured in movies like "Jaws" is immediately counteracted by three small, shallow pools housing the majority of the Aquarium's shark population. Most of these sharks grow only three feet in length and tend to huddle together, giving the appearance of "cuddling." In contrast to the great beasts of film and legend swimming about in the depths of the Pelagic zone of the ocean who sometimes surface to stalk humans, these small sharks dine on sustainably-caught, restaurant-quality seafood graciously given to them by their human caretakers, Aquarium staff. These more diminutive, safe-to-touch creatures, rather than the larger sharks housed in enclosures behind the shallow pools, largely serve as the Aquarium's archetype for re-educating the public about the beast whom the public appears to misunderstand (Dobson, 2011; Panoch & Pearson, 2017).

The Shark Lagoon, devoted to experiencing, touching, and learning about sharks, is not the only location where people feel the predator's presence. Bonnethead and zebra sharks, for example, roam the waters of the Tropical Pacific Gallery upstairs and leopard sharks the Blue Cavern exhibit downstairs. In the So Cal/Baja Gallery, visitors can glimpse an unborn shark alive inside an egg case as staff have removed a small piece of the case and replaced it with a clear plastic "window" for display. The Aquarium's two gift stores feature sharks in games, children's books, toys, and jewelry. Periodically, the Aquarium also invites guest lecturers who discuss the plight of the shark, its role as an apex predator in ecosystems, and shark conservation.

In this chapter I show how the Aquarium's various sites that teach about sharks produce a hidden curriculum (Lloro-Bidart & Russell, 2017; Lukasik, 2013; Pedersen, 2004) comprised of three related discourses, all of which rely on naturalistic ethics (Rolston, 2002). These discourses include: (1) a focus on the shark's role as apex predator of *nonhumans* to perform ecosystem services; (2) an emphasis on the mischaracterization of the shark as a killer of *humans* in the media; and (3) a re-imagining of the shark as cute, cuddly, and desiring amiable relations with other sharks and humans. Together with material and tactual experiences involving live sharks, these discourses construct what I call the "disembodied shark." This largely misunderstood shark, who is not materially present at the Aquarium or in the oceans, serves as a kind of conservation flagship species (Douglas & Verissimo, 2013) worthy of human care and concern. The Aquarium does not funnel its efforts into reconstructing the image of any one particular shark, like the great white; rather it seeks to construct *all* sharks as benevolent and worth saving.

The pedagogical assumption undergirding these practices is that visitors to zoos and aquariums can meet and interact with, as well as learn about ambassador or flagship species and subsequently come to care more about ecosystems, habitats, and their wild counterparts. Some conservation educators and educational researchers seem to have uncritically accepted such a view, despite conflicting evidence supporting these assumptions (Marino, Lilienfeld, Malamud, Nobis, & Broglio, 2010; Malamud, 2017). This chapter, therefore, works to complicate and, in some instances, unsettle these assumptions.

Ecocentric Philosophical Narratives

At the Aquarium, the disembodied shark is posed as an important apex predator of only *nonhuman* species, as it is disinterested in attacking or eating *humans*. The demarcation between shark-as-*nonhuman* killer and shark-as-*human* killer is critical as it authorizes staff to describe sharks in naturalistic terms, i.e., as beneficial for ecosystem stability and benevolent toward *humans*. The sharks' role as apex predator is highlighted to justify its importance to the marine biosphere. Staff do not emphasize individual animal rights but rather echo an ecocentric philosophical position by focusing on their role in maintaining ecological stability for the good of the biosphere. Aldo Leopold, wildlife manager, forester, and environmental philosopher, succinctly describes such a view, "Yet these creatures are members of the biotic community, and if (as I believe) its stability depends on its integrity, they are entitled to continuance" (1949, pp. 246–247). Sharks, therefore, have "relative value" (Callicott, 1980, pp. 324–325) as they perform important ecological roles and services that ultimately also benefit us.

In what I call naturalistic "aesthetic nature renovation," staff (along with the sharks) also work to connect sharks' perceived aesthetic value to their intrinsic value in an attempt to revitalize the image of the shark, where intrinsic value is defined as, "what it [the animal or aspect of nature] is in itself, and not simply instrumentally, for the pleasure it brings us" (Rolston, 2002, p. 137). According to ecocentric philosopher, Holmes Rolston, two kinds of aesthetic qualities exist: aesthetic capacities (beholders' capacities for experience) and aesthetic properties (those that objectively lie in natural things). When humans come to aesthetically value the natural world, then, they are not only "excited by the incoming data … translate[ed] as aesthetic value" but "value the processes and products of generative nature that we are discovering" (Rolston, 2002, pp. 132–133). To illustrate, he explains how humans appreciate the "grace" in the motions of a leaping impala (i.e., their aesthetic capacity), but for this grace to be realized, impalas rely on "muscular power" derived from "evolutionary achievement" (i.e., their aesthetic property) (p. 133). This kind of aesthetic valuing, which the Aquarium pedagogically embraces, is designed to incite an appreciation for the beauty humans perceive *and* the natural processes that produce that beauty, leading to intrinsic valuing and appreciation of the biotic community as a whole. Sharks at the Aquarium, therefore, become "cute," "cuddly," and "athletic" and therefore deserving of human care and concern.

Biopolitics and Speciesism

In this chapter, I return to the Foucauldian and neo-Foucauldian theories (Darier, 1999a, 1999b; Foucault, 1978; 2003; Smith, 2001) discussed in Chapter One to problematize the Aquarium's strategies for promoting shark conservation, which ultimately promote a hidden curriculum that commodifies individual shark lives and teaches people that sharks have value insofar as they are valuable to us. Simultaneously anthropocentric and ecocentric, the institution engages in a kind of biopolitical speciesism, which in some aspects parallels Michel Foucault's (1978, 2003) biopolitical racism. For Foucault, biopolitics confers to the state (or institution) the power to discipline and regulate life; these practices authorize the death of some beings (those deemed biologically inferior) in order to "make live." Like with the lorikeets discussed in Chapter Four, this biopolitical regulation is neoliberal since shark edutainment helps to fiscally sustain the institution, and it is disciplinary insofar as the educational programming is designed to foster certain kinds of shark life.

Yet the bodies of the sharks captive at the Aquarium are not only aestheticized in order to save the disembodied shark (constructed as biologically superior, as it possesses all of sharks' "positive" qualities, but none of the "negative" ones); they are also subjected to a "political death" as they are relegated to the category of biologically inferior and ostracized from life in the wild (Foucault, 2003). Thus, while such discourses and tactual encounters seemingly promote care and empathy for all sharks while at the same time contributing to the fiscal stability of the institution, a problematic hidden curriculum—explicitly connected to the humanism of conservation education practice and research—teaches otherwise. In the concluding chapter of this book I argue that if zoos and aquariums continue to capture and biopolitically regulate nonhuman bodies in the service of conservation ends, then part of the public conservation education project should embed knowledge of not only the violence humans inflict on animals in the wild, but also that which is perpetrated against wild animals in captivity.

Shark Lagoon Exhibit Background and Description

The Shark Lagoon features two main sections: three shallow, interconnected touch pools housing the Aquarium's small (three feet or less) sharks and one massive tank the great sharks inhabit. Above these pools hangs a large, turquoise blue canvas banner that reads, "SHARK LAGOON." The turquoise banner is complemented in color by canvas turquoise tarps that cover and shade, depending on the time of day, the entire exhibit. (See Figure 5.1.) All of the pools are surrounded

Figure 5.1: The Aquarium's Shark Lagoon Exhibit; a crowded day in July 2012—

by rocky formations similar to a high-end in-ground pool, designed to make the Lagoon appear more natural. The center pool and the far left pool are interconnected and at any given time of day, except during feeding, visitors can touch the stingrays (Pacific cownose ray and reticulate whip ray) and a medley of full-grown female sharks (epaulette, white-spotted bamboo, brown-banded bamboo, and zebra sharks) who inhabit these pools. Because the scavenging tropical fish cohabitating with the sharks and rays have fragile skin and gills, the Aquarium forbids visitors from touching them. Staff on the microphone regularly remind visitors that the tropical fish are off limits.

The sharks tend to nestle close to each other, making it difficult, but not impossible, for visitors standing on the opposite side of the pool to interact with them. Staff, although not entirely sure why the sharks do this, theorize that much like schooling fish, it makes a group of small animals appear larger to predatory species. Other staff describe them as "napping" when they exhibit this behavior that is only interrupted during one of the feedings scheduled throughout the day or when staff gently lift a shark to the surface for humans to have body-to-body contact. At the time of feeding, staff announce a "no touch time," forcefully encouraging visitors to remove their hands from the water to avoid being accidentally bitten by a hungry shark anxious to gobble up the bits of restaurant-quality seafood staff fling into the water.

To the right of these two interconnected pools is a smaller pool that houses baby bonnethead sharks approximately one foot in length, tropical fish who act as cleaners of the pool, horseshoe crabs, and various conchs. As obligate ram ventilators, the baby bonnethead sharks, unlike the epaulette, bamboo, and zebra sharks, must constantly swim in order to flush vital oxygen into their gills to breathe. While permitted to do so, visitors often find baby bonnetheads difficult to touch due to their continual motion. These sharks, unlike the smaller sharks, cannot reside their entire lives in these pools because they can reach up to five feet in length. Visitors may also touch the conchs and horseshoe crabs, but not the tropical fish.

Directly behind these shallow pools the Aquarium houses its larger sharks (a blacktip reef shark, a freshwater sawfish, a sand tiger shark, a sandbar shark, a whitetip reef shark, and a zebra shark at the time of my fieldwork) in a huge tank that resembles an in-ground swimming pool. Visitors are able to view these sharks through two means: (1) behind the far right shallow pool through a top-open area; or (2) to the left of the far left shallow pool through plexiglass in a down-sloping viewing area. Visitors may "interact" with some of these sharks during feeding time as they watch Husbandry staff give a live "show" or if they purchase a "Shark Encounter" experience that allows them the opportunity, while being supervised by a staff member, to feed one of the larger sharks. They are not permitted, however, to touch the larger sharks in the same manner they would touch the smaller sharks housed in the shallow pools.

Staffing the Shark Lagoon

The Shark Lagoon is mostly staffed throughout the day with volunteer educators.[1] The number of staff watching over the Lagoon fluctuates based on the number of volunteers available as well as the number of visitors expected (the Aquarium produces "expected visitorship" figures each day). During the 14 months I collected data there, I witnessed days with only one or two staff members and a handful of visitors, and other days with four or five staff members and hundreds, if not thousands of visitors. A small staff area called the "Volcano," located directly behind the center shallow pool, serves as the site where volunteer staff members stand or sit when they talk on the microphone to visitors about the various animals that inhabit the Lagoon. During volunteer training, paid staff teach the volunteers about sharks, many of whom have little to no background in either education or marine biology. Unlike many of the other large exhibits at the Aquarium, staffers speaking at the Shark Lagoon do not have a script and which shark facts volunteers share with visitors and in what order is generally up to the individual staff member. Examples of these facts are depicted in Table 5.1. All staff on the microphone verbally remind visitors that they are to touch the sharks with only two fingers (since one finger is a "poke" and a whole hand is a "grab") and most visually demonstrate the proper method by holding out two fingers as they make a swaying motion in the air to simulate petting an animal. Visitors encounter the "two-finger" touch rule at every exhibit allowing physical interaction, except the Lorikeet Forest where visitors are advised not to touch the birds at all.

Some sharks are egg-layers and some give live birth. The sharks in the shallow pools available for touching are egg-layers. These egg cases are called "mermaids' purses."
Sharks have rough skin called dermal denticles.
Most sharks in the world are smaller than three feet in size, despite popular conceptions that they are all the size of great white sharks. Eighty percent of sharks are less than eight feet in length.
Some sharks, called "obligate ram ventilators," have to swim constantly in order to breathe.
Sharks are finned in some parts of the world, which is contributing to their demise. Staff are instructed in their volunteer handbook, however, not to mention the specific countries that fin sharks and consume their flesh, but this rule is occasionally broken.
Sharks have skeletons made of cartilage, so almost no fossils of sharks exist.

Table 5.1: Shark facts staff share at various exhibits.

In addition to the person on the microphone staffing the Volcano, another staff member may "wade" in the center pool while others may wander around to chat with visitors and monitor their behavior, and another might stand at the Shark Cart that is typically stationed behind the far right shallow pool, although it can also be found indoors on some occasions. The staff member "wading" dons the typical Aquarium uniform (navy blue shirt and khaki pants or shorts) along with rain boots, since they are partially submerged in the warm, salty water with the sharks. The wader primarily teaches visitors about the sharks and lifts sharks to the surface of the water so visitors who cannot reach deep in the pool can touch them. Visitors in wheelchairs or small children, for example, typically cannot stretch their arms far enough into the water to touch the sharks. The staff member working the Shark Cart chats with visitors and allows them to touch shark teeth and shark jaws, most of which are plastic replicas.

"Big Trop" and Blue Cavern Exhibit Descriptions

"Big Trop," staff's colloquial term for the gigantic tank, is the Aquarium's largest exhibit. Located in the Tropical Pacific Galley upstairs, visitors can view exhibit animals from three locations or "windows" inside. It holds 350,000 gallons of water and houses over a thousand animals, including fish, corals, sea turtles, and zebra and blacktip reef sharks. The exhibit, modeled after the "Blue Corner" off the coast of the country of Palau, is a replica of a well-known dive site. Several times a day paid interpretation staff (on the "dry side") and volunteer dive staff (on the "wet side") give 10 to 15-minute shows at the exhibit's largest window. The paid staff doing these presentations are usually entry-level and part-time.

The Blue Cavern Exhibit, located downstairs in the Great Hall, is modeled after Santa Catalina Island's Blue Cavern Point, a kelp forest ecosystem off the coast of Southern California. The exhibit, spanning three stories, contains 142,000 gallons of water and houses leopard sharks, California moray eels, California sheephead, giant sea bass, garibaldi, and other organisms. Like the Big Trop exhibit, shows occur here several times per day as a paid presentation staff member and volunteer dive staff member talk for approximately 10 to 15 minutes.

During both of the shows, dive staff wear special underwater microphones along with typical diving accoutrement. Their voices are audible, but often muffled as staff have to coordinate breathing and speaking. Unlike in the Shark Lagoon, both shows are scripted, although presenters and dive staff often ad lib as they highlight stories and facts they consider particularly salient or interesting. During the busy summer season, staff give more shows per day to accommodate the increased number of visitors.

The Disembodied Shark

Apex Predator of Non *humans*

Sharks, generally considered apex predators in their respective ecosystems, stabilize prey populations and maintain ecological balance. When their prey flourish unchecked, they often unleash drastic and far-reaching consequences. For example, great shark population declines on the United States' eastern seaboard between Cape Cod, Massachusetts and Cape Canaveral, Florida, have caused mesopredator populations, particularly cownose rays, to explode and destroy bivalves and habitat (Myers, Baum, Shepherd, Powers, & Peterson, 2007). Scientists believe, therefore, that sharks perform important ecosystem services by stabilizing food webs that ultimately benefit humans.[2] In *The Land Ethic*, Leopold illustrates the ensuing mayhem when apex predator populations decline: "Domesticated species from other lands are substituted for wild ones, and wild ones are moved to new habitats. In this world-wide pooling of faunas and floras, some species get out of bounds as pests and diseases, others are extinguished" (1949, p. 254). The concern about the sharks' ecological importance, predicated on conservation science and ecocentric philosophy, thus spurs discourses at the Aquarium that ground the shark's value in scientific holism, "the interrelatedness of organisms in natural systems and the connections between those systems" and axiological holism that "ascribes value [to an organism] on the basis of usefulness to the biosphere" (Marietta, 1995, pp. 55–58).

During a "Blue Cavern" dive show at the Aquarium, for example, an interpreter constructs sharks as valuable due to their role as apex predators, "I think they [leopard sharks] are so important to the open ocean as apex predators. They keep the ecosystem in balance by killing the sick and dying animals." Similarly, a placard flanking the front side of the Shark Lagoon entitled "Why are Sharks Important" explains:

> *Sharks are apex predators, or predators at the top of the food chain. These apex predators play an important role in ensuring healthy and diverse populations by feeding upon weak and diseased marine animals. When apex predators such as sharks are removed from an ecosystem, an imbalance can result. An unbalanced ecosystem can lead to population explosions, disease and shortages in food sources.*

Resonating with ecocentric philosophy (Callicott, 1980, 1989; Leopold, 1949), this discourse shifts the conversation from the perceived savagery of individual sharks to the disembodied shark's overall role in maintaining ecosystem balance, an extrinsic value. However, these statements disrupt two key messages the Aquarium promotes, highlighting the contradictions present when staff must embrace pedagogical practices that present animals as appealing to humans. First, staff at the exhibit site frequently highlight how sharks *do not* consume the other fish in the exhibit, suggesting they are not predators at all. And second, staff often emphasize that Aquarium sharks' do not view humans as food simply because of the method utilized to feed them—not because sharks never view humans as food. The Big "Trop Dive" Show excerpt depicted below demonstrates these contradictions:

> *Jeff: Our sharks do not associate our divers with food. They don't see the fish in there as food. We actually feed them above and there's a big plastic piece that we go under water. The sharks know when that goes under the water they are going to get fed. And they are target fed from above.*
>
> *[The diver then mentions how the sharks are target fed.]*
>
> *Jeff: Yeah, with tongs. Don't forget about the tongs. It's very important.*

In the excerpt, Jeff strips the shark from its role as apex predator, which directly impugns one of the Aquarium's most prominent reconstructive shark discourses. He does this to counteract the image of the bloodthirsty shark the Aquarium presumes many visitors expect, but in so doing, diminishes the extrinsic value of the shark the Aquarium works to establish through accentuating shark as an apex predator.

Jeff also underscores the importance of sharks not associating exhibit divers with food. As I discuss below, one of the dominant Aquarium shark discourses involves convincing visitors that sharks *absolutely* do not desire to consume humans, that attacks only occur in cases of mistaken identity. (See Figure 5.2.) Further, staff emphasize that sharks do not find the taste of human flesh palatable. As Jeff explains the feeding process to visitors above, he implies that Aquarium sharks do not wish to eat their human diving companions because of the method utilized during feeding, not because sharks are implicitly disinterested in eating humans. This rhetoric implies that sharks *can* attack and harm people, despite the prevailing message that they do not.

Figure 5.2: Diagram staff utilize at Shark Lagoon and the Shark Cart to demonstrate to visitors how sharks attack people through "mistaken identity"

A few staff members recognize these contradictions and subtly resist the Aquarium's mainstream discourses. Although the Big Trop Dive Show is scripted, interpreters have the opportunity to infuse bits of their own talk, as Fred illustrates below while conversing with the diver, Vic:

> *Fred: Well, you know, Vic, there is so much to do about sharks and things like that and people are very fearful of sharks. And in reality, we should all be, maybe a little fearful of things that we don't know that much about. But the reality is that you have a far greater chance of maybe having a vending machine fall on you than you ever do of being attacked by a shark. But, you know, we want to always caution people. If you encounter an animal that you are not familiar with, whether it's out on land or out in the ocean, the best and safest thing to do is simply get out of the water for a little while and get to a safe area. Or, just remain nice and calm and still. Very much like what Vic is doing. You'll see these sharks going all around and underneath Vic and past Vic, but you don't have any qualms about being in there at all, do you Vic?*

Vic confirms that he feels safe in the tank because the sharks are fed with poles and do not associate people with food. Fred acknowledges the threat of shark attack is rare through comparison ("people have a greater chance of being hit on the head with a vending machine"), then slightly retracts his statement, suggesting that should visitors encounter a shark in the ocean, they ought to exercise caution. Fred, the only Aquarium interpreter I observed advancing such admonitions, perhaps because of his own professional background working in companion animal shelters, suggests that sharks are not the totally innocuous beasts mainstream Aquarium rhetoric purports. Nature writer David Quammen (2003) asserts in his chronicle of "man-eating" beasts, *Monster of God*, that downplaying the threat or seriousness of animal attack comes more easily to those who do not regularly co-exist with flesh-eating animals, a point echoed in the online shark bite support organization "Bite Club" that is populated by many male surfers who have experienced shark bites (Appleby & Pennycock, 2017). Humans living with large predators by choice or necessity must therefore confront the perpetual fear that they or family members may be attacked or eaten.[3] Fred, having decades of experience working directly with animals, is more trepidatious than some of his colleagues who sometimes have had little to no real-life experience with wild animals.

"Mistaken Identity"

As discussed in the preceding section, the shark is constructed as an important apex predator in marine ecosystems. An integral aspect of this discourse is the shark as predator of *nonhuman* species. While popular media, in movies like "Jaws" and in documentaries like "Shark Week," construct sharks as *human*-eating, the Aquarium envisions a benevolent *flesh*-eating beast who maintains ecosystem balance and mistakenly attacks humans on rare occasions, mainly due to poor vision or in defense.[4] The passage below, a typical conversation Aquarium staff members have with visitors over the microphone at Shark Lagoon or at the Shark Cart, illustrates how staff construct shark attacks as cases of mistaken identity. While visitors interact with and learn about *small* sharks, in the excerpt below, Rose seizes the opportunity to reconstruct the great white shark, whose perceived viciousness is not attributed to maleficence but an acquired evolutionary trait—poor vision causing great sharks to mistake humans for seals or sea lions:

Rose: The sharks in here are not babies ... These are full grown bamboo sharks. Contrary to popular Hollywood dimensions, this is the average size of 60 to 70 percent of all sharks in the world. We've all been taught to think that they are like great whites and that they are all hunting us, which they are not. This is the average size of them ... You see this one? The one that is beautiful that is coming toward you? That is an epaulette shark ... I want to show you guys a picture ...

Rose goes over to the Shark Cart and grabs the laminated cartoon drawing of a person in a wetsuit on a surfboard who, to a shark spying the person from below, appears seal-like. [See Figure 5.2.]

Rose: We get a lot of questions here at Shark Lagoon about shark attacks, especially white shark attacks. So somebody brought this to have us help and have you understand why that would happen. This is what a great white shark would see under the water as it's looking up. It would see a seal, which is what it likes to eat. And a seal is black and from underneath it is going to look black to him. Next to that is a surfboard, with a human being on the surfboard. And the human being is probably wearing a wet suit, which is black. And if you were to put his arms down like that, as he's going like that through the water, that's why there is a mistake made. Great whites do not purposely go after human beings. In fact, it's been proven that if they take one bite, they know they didn't get a seal; they're not going to do anything else to you. That's the good news. The bad news, usually they bite you badly enough, you know, it's not very pretty. But, it's not on purpose. So you can see the difference on that ...

In this passage Rose asks visitors to challenge popular assumptions about sharks such as their large size and their viciousness. Visitors are encouraged to do this through "becoming" (Haraway, 2008; Pedersen, 2011) shark for a brief moment. As they touch the small sharks, staff ask visitors to imagine life as a visually challenged, famished shark gazing upon what appears to be a sea lion, seal, or sea turtle. (See Figure 5.2.) In this multispecies entanglement, humans have the opportunity to re-imagine their relationship with the *disembodied shark* the Aquarium constructs through a touch experience with an *individual* shark. Akin to Donna Haraway's (2008) assertion that we should "look back" at animals as we experience relationships with them, the Aquarium encourages visitors to not only view sharks, but also to *touch* them as they imagine what life might be like for sharks in the rough waters of the open ocean.

Staff explicitly describe the touch experience as one that facilitates the construction of the shark the Aquarium imagines for visitors—an innocuous, friendly, and mischaracterized beast. In the following interview excerpt, as I chatted with Helen, a paid staff member, about what visitors do at exhibits, she segued into this discussion about touching:

> *Touching is a great a way to, you know ... a great way to talk to people. Especially, I love the people who are afraid to touch. Encouraging them and getting them to touch is something I love to do [laughing]. Parents love it. It's a great way to uh, it's got a number of benefits. You know, you have encouraged somebody to experience their world, break down myths and misconceptions, especially with respect to sharks.*

The focus on the touch experience of the *human* is directly linked to "breaking down myths and misconceptions" because it allows the interpreter to dialogue with visitors as they have positive, embodied experiences with sharks.

To further convince visitors of the shark's disinterest in harming humans, staff members physically interact with both the small and large sharks, demonstrating to visitors their own attempt to "become with" (Haraway, 2008). They wade with the sharks in the shallow pools at the Lagoon, partly to lift sharks to the surface so that people who cannot reach into the exhibit may touch them, and partly to re-emphasize the sharks' harmlessness. For example, when asked to elaborate those messages she believes are most important to convey to visitors, volunteer Education staff member, Sarah, commented:

> *You know, people view sharks as this menacing, man-eating beast that they really aren't, you know? And I'm walking in there with a hundred little sharks and you can pet them and they're just sweet and docile and they just want to live their little lives just like everybody else on the planet.*

While volunteer staff members like Sarah wade with the small sharks in the touch pools, highlighting their aesthetic capacities of "sweetness" and "docility," dive staff members actually swim in the large tanks with sharks. During shows, they converse with paid Education staff on the "dry side." The excerpt below illustrates typical interactions between Dive staff, Education staff, and animals during the Tropical Dive show. During the show, staff not only swim with bonnethead, zebra, and black-tip reef sharks, but villainize the golden trevally, a harmless-looking fish, to distract visitors from sharks:[5]

Jeff: These little bonnetheads ... this is as large as they get. A lot of people think that Hammerheads are aggressive sharks. This is not the case ... We talked about how they are not really too aggressive, but there is a group in there that is very aggressive when it comes time to feed and that is going to be, one of them is hangin' out right down there [points to where they are in the tank]. The rest of the motorcycle gang is going to be coming later on. That's the golden trevally.

Isabel, the diver, points out a golden trevally approaching her from behind.

Jeff: There we go. Wow, right there. The ones with the black, vertical stripes on them. They also have that little yellow trim ... And why do I call them the motorcycle gang? Why do we call these ones the motorcycle gang?

Isabel explains that the fish are "ornery" and calls them the "motorcycle gang" because, as schooling fish, they prefer to swim together, like a motorcycle gang travels together. She tells the crowd that when the golden trevallies eat, their stripes get much darker, signaling to other fish it is time to eat. She also shares that they are super aggressive feeders and that the diver who target feeds them does it with a bucket, attempting to make the feed last as long as possible so that the other animals in the exhibit can eat. If the trevallies are not eating, they will bug the other fish while they attempt to eat.

The dialogue above further evinces the Aquarium's arguments about mistaken identity. Not only are visitors incorrect in assuming shark malevolence, but visitors err in believing harmless appearances signify innocuous behavior. The image of the "motorcycle gang" confiscating all of the food flung into the large tank conjures up media images of actual motorcycle gangs seizing neighborhoods or towns for their use. This caricature, coupled with divers swimming with the great sharks, re-focuses visitor attention on the "violence" of the golden trevallies and not on the perceived viciousness of sharks. Instead, sharks become the respectable "sheriffs" who benevolently maintain ecosystem balance for the good of the many.

Aesthetic Nature Renovation

Although the Aquarium is not a conservation group per se, its mission statement, carefully crafted to align with the Association of Zoos and Aquariums (AZA) guidelines, makes it clear that "conserving natural capital" is a priority. Following Kathryn Yusoff, I focus on the Aquarium's discourses as a form of "aesthetic practices" as I ask: "How do beings come into being?" (2011, pp. 579–580) and, more specifically, how does the Aquarium's disembodied shark come into being? (also see Fawcett, 2013; Lloro-Bidart, 2018; Russell, 2005). Aquarium sharks, whose bodies are biopolitically regulated, and Aquarium staff members actively participate in what I call, drawing on Rolston (2002), "aesthetic nature renovation."[6] Staff do this by assembling a hodgepodge of attractive features—cuteness, athletic grace, toy-likeness—and eschewing negative features—the ability to shred the flesh of other organisms, including humans—thereby fashioning a beast worthy of saving.[7] These processes, "aesthetic" because they explicitly induce visitors to value sharks through both aesthetic capacities and aesthetic properties, are also "renovative" in that they seek to expunge visitors' prior conceptions of sharks while simultaneously imagining new ones. The small sharks, as an integral part of this assemblage, actively participate in re-inventing the image of the shark by allowing humans to touch them.

"They cuddle" and "they're like wind-up toys." The bodies of the small sharks in the shallow pools, as a form of Foucauldian biopower (Foucault, 1978, 2010; Luke, 2000a, 2000b; Rabinow & Rose, 2006), serve as the archetypes for the "cute," "cuddly," and "small" shark the Aquarium constructs for visitors as staff and sharks work to entreat human aesthetic capacities through highlighting shark social and physical aesthetic properties. Visitors are permitted to gaze upon *and* touch these sharks due to the exhibit physical structure that confines sharks and allows visitors to approach from virtually any side of each of the three pools, except the back of the exhibit. The small sharks tend to gather in particular sections of the pool, which the staff often describe as "cuddling," "huddling," or "piling up" as sharks "nap" or work to "appear larger to a predator," although scientists are not in fact sure why they behave in this manner (Bauernfeind, 2017). In the excerpt below Candy, an Education staff member, describes the sharks' behavior to visitors at the Shark Lagoon during a live feeding, where bits of restaurant-quality seafood are tossed into the exhibit:

> *And they actually huddle along the walls like that for … safety purposes. In the wild, I mean a three-foot shark isn't very big in the wild. So to protect themselves, they*

> *huddle just like that so that predators actually think that it's a bigger animal, except just one small shark. This is how they protect themselves from predators so that they live longer ... So they do this all the time. We say that they're napping. We say that they're hugging each other. But it's basically just so that they can keep each other safe.*

Mike, an Education staff member, similarly describes the sharks' benevolent sociality during feeding time in the pools:

> *There have never been fights over the food [during a scatter feeding]. There's more than enough ... But they like to cuddle and they get together in different parts of the Lagoon. And the most current theory is they do this to appear larger to a predator than they are. If the predator attacks anyway, they scatter in a billion directions, give the predator a heart attack, and they get away.*

These passages illustrate how Aquarium staff, together with the sharks, construct a beneficent "shark" as they appeal to human aesthetic capacities—sharks "cuddle" or "hug," making them attractive or aesthetically pleasing to watch.

However, cuddling and hugging are also social aesthetic properties because: (1) sharks' social relationships with each other are "objectively there" whether or not humans are present; and (2) sociality is an evolutionarily developed trait, much like musculature or color. While the full extent of the social relationships may be generated by humans desiring sociality in sharks (particularly with respect to sharks' desires to socialize with humans), animal behaviorists and biologists have established that many species of sharks are objectively social, i.e., social regardless of human presence. Sharks demonstrate these behaviors in an objectively real sense as they have evolved to be social, in the absence of human interpretation. This connection between aesthetic capacities and aesthetic properties, according to Rolston, is critical to translating aesthetic valuing into intrinsic valuing: "This is ecological aesthetics, and ecology is vital relationships ...This 'interest' does lead me to care about its integrity, stability and beauty" (2002, p. 139).

Thus, amiable shark-shark relations and newfound knowledge that sharks actually do not desire human flesh, coupled with visitors' ability to touch sharks to create the *possibility* of human-shark social relationships for visitors at the Lagoon, characterize a shark worthy of being saved.[8] Candy illustrates these possibilities in the interview passage below, as she responded to my question about teaching at the Aquarium:

> *So the sharks we have in Shark Lagoon. When we tell them it's okay to touch, they're, like, really? Why? They're cool, you know? We're in there all the time when we're wading in the water. They don't come and bite us because we're in there. They'll come and snuggle and try to figure out what our legs are doing in there [when staff wade in the small pools with the small sharks]. But, there's never been a shark that would purposely come up to us and bite us. So, I think a lot of that is trying to get the visitors to understand that these animals are safe. And the purpose of the Aquarium is to show people that animals need our help more than anything. Instead of us trying to kill them because they look scary. That's the majority of what I try to tell them when I'm up there.*

Candy simultaneously describes a shark safe to touch ("They don't come and bite us …") and one interested in relations with humans ("They'll come and snuggle and try to figure out what our legs are doing in there"). Here, she draws on aesthetic capacities; humans at the Shark Lagoon desire social relationships with the small sharks who appear to reciprocate, thereby aestheticizing the small sharks.

Following Rolston (2002), however, sharks' apparent eagerness to socialize with humans is *not* an aesthetic property; one cannot be certain interspecies sociality (particularly with humans) is an evolutionary trait objectively there without human presence. In Rolston's (2002) view, translating shark *intra*species sociality from an aesthetic property (i.e., evolution has bestowed intraspecies sociality upon sharks) to an aesthetic capacity (i.e., humans find watching these relationships aesthetically pleasing) could potentially lead visitors to locate intrinsic value in sharks. Shark *inter*species sociality (i.e., desiring social relationships with humans) certainly appeals to aesthetic capacities, but not on account of any aesthetic property because a shark's desire to socialize with humans is not something objectively there, in a shark, prior to human contact and interpretation. It is questionable, therefore, if these arguments can lead to intrinsic valuing. The implications of this flaw are important; the Aquarium clearly aims to inspire people to care more about the plight of sharks. Recall that Candy emphasizes that sharks *deserve* human assistance ("The purpose of the Aquarium is to show people that animals need our help …") precisely because they are docile creatures desiring human interaction. While aesthetic valuing matters in an artistic sense, "it may disorient us and leave us with too weak a locus of value to protect all the values in jeopardy" due to its strict focus on human interests (Rolston, 2002, p. 128). Further, in contriving an Other fit only for human consumption (Smith, 2001), it "assign[s] a very limited conception of biotic citizenship" to them (Yusoff, 2011, p. 586).

The Aquarium not only capitalizes on discourses of sharks "cuddling," "snuggling," and pursuing relations with humans, but also describes how they manifest "cuteness" as they appear like "wind-up toys" and "play tag" (particularly the baby bonnetheads). For example, Sarah, on the microphone at the Shark Lagoon in the passage below, describes their "toy-like" appearance as she converses with visitors. Shortly after, Rose highlights their locomotion as they "play tag":

> *Sarah: There's one baby nurse shark in here. The last I saw he was hiding out over there … and there are two baby bonnethead sharks in our little Lagoon over here … They just put them in there and they're super cute. They're just getting used to being out here. So, they look kind of like little wind-up toys swimming around.*
>
> *A few moments later …*
>
> *Rose: But they're, did you get to see our little baby bonnetheads? They just brought them in yesterday. There's one that's going to be coming along. That's a bonnethead shark. Two babies … There they go. They're playing tag. Yeah. Aren't they cute? I think they will lighten up afterwards. A lot of them are more of a grayish color. They are very, very, very cute. We're all so excited.*

Like Rolston's (2002) "graceful" impala, beautiful because of its motions that please humans (its aesthetic capacities) and its musculature (its aesthetic property), sharks are touted as "cute" and like "wind up toys," suggesting athleticism, grace, locomotion, virtuosity, speed, and an ability to entertain. Sharks' small, streamlined bodies and fins, aesthetic properties insofar as they are evolutionary achievements that invoke a human appreciation for sharks and their role in the biotic community, ignite aesthetic capacities (Rolston, 2002). Due to the connection between "art" and "biology" or aesthetic properties and aesthetic capacities, these discourses in Rolston's view could incite intrinsic valuing, although I will problematize that idea in the concluding sections of this chapter.

"I Want a Nice One That's Going to Have Puppy Dog Eyes"

The tensions between the shark the Aquarium seeks to construct for visitors and the popular media image of the shark also unfold in the Aquarium's gift shop, Pacific Collections. Figure 5.3 displays two plush toy versions of the shark visitors may purchase, one more closely resembling a dolphin (doe-eyed, tucked-under teeth inside a small mouth, and a puffy head and body) and the other a somewhat more accurate version of a shark (eyes set more on the sides of the head, visible teeth, and streamlined face and body).

Figure 5.3: Shark plush toys available for purchase in Pacific Collections

While the shop provides visitors a few non-living versions of cute, cuddly, and toy-like sharks for a price (consuming to conserve at the Aquarium extends beyond purchase of the entrance fee), interestingly, the majority of offerings portray the typical media version of the shark with words like "attack," "monster," and "angry", conjuring visions of human killers. (See Figure 5.4.)

Figure 5.4: Typical shark items for sale in Pacific Collections

Some staff, like Maria, recognize that the images on display in the gift shop contradict the shark that the Aquarium works to re-imagine. As a new Education volunteer at the time of our interview, she expressed concern about these discrepancies:

> *Maria: I feel they [visitors] come in, especially with the sharks, in an ugly way to see them and then I feel it's our job to make them see the other way. They get to choose which ones they believe, but it's our job to show them. Look, they're not so bad. They're not so scary like the movies portray.*
>
> *Teresa: And that's particularly with the sharks, not necessarily with everything else?*
>
> *Maria: Especially with the sea lions. They're like, "Oh they're so cute. I want to touch them." But, I mean, you can see their giant teeth and they would most likely bite you rather than a shark that you're touching. He's going to be—he would get away from you—rather than bite you.*
>
> *Teresa: So people think that the sharks are going to harm you and they're not and they think that the sea lions are huggable, and they're not?*
>
> *Maria: Yes. [laughing]*
>
> *Teresa: How do you think it is that people develop those perceptions?*
>
> *Maria: TV. I mean, I used to think that, too. And it's just TV. You see cuddly little creatures and if you look at the shop itself you see cuddly little animals of sea lions and seals and then the sharks, they show the teeth and they make it look really vicious.*
>
> *Teresa: I see …*

> *Maria: Some of the sharks are cute. Other ones you can just see all the teeth in them. And kids are going to be like, I don't want that. I want a nice one that's going to have puppy dog eyes and everything.*

Maria's concerns depict a real affront to the Aquarium's goals of aestheticizing sharks into a creature all humans would want to love. Not only do staff confront contradictions as they chat with visitors expecting to encounter the beast of "Jaws," but they also have to cope with contradictions within the institution's material offerings to visitors, such as toys, games, and books. Particularly interesting in this passage is Maria's comparison of sharks to typical charismatic megafauna like sea lions. Like the staff member, Jeff who caricatures golden trevallies as "motorcycle gangsters" to show that innocuous appearances do not always equate with benevolence, Maria points to the viciousness of seal and sea lions, who appear cute, but "would most likely bite you." What both of their discourses reveal are the inherent difficulties and inconsistencies in aestheticizing any aspect of "nature," which I explore in the concluding sections of this chapter and in the closing chapter of the book.

Speciesism and Sharks' Political Death

Despite the Aquarium's attempts to aestheticize sharks, they remain predators indebted to the regular consumption of flesh for their own survival, and while the size may vary, all sharks possess razor-sharp teeth evolved for this task. In the Aquarium's admirable effort to reimagine a shark deserving of human care and concern, staff carefully evade these characteristics, accentuating what they believe will appeal to human sensibilities. Rolston asserts that this kind of renovation is customary among artists, "The aesthetician repairs nature before admiring it. Landscape artists and architects are like flower arrangers; nature does provide raw materials, but raw nature is quite aesthetically mixed" (2002, p. 136). He goes on later, however, to challenge the efficacy of this approach in conservation discourses. Utilizing the crane, "an ungainly hulk" not meeting the "pretty criteria" as an example, Rolston (2002) contends that humans cannot value the crane based on artistic aesthetic criteria alone; rather, they must develop "a sense of biotic community" with an organism whose evolutionary accomplishments (survival throughout the last 40 million years) remain, on some level, aesthetic. More importantly, valuing what is biologically rather than artistically aesthetic about the crane, suggests that humans can come to intrinsically value what might typically be considered non-aesthetically pleasing aspects of nature, in much the same way some people come to value other unloved creatures, like insects (e.g., Boileau & Russell, 2018; Lemelin, Harper, Dampier, Bowles, & Balika, 2016).

In his edited volume, *Discourses of the Environment*, Eric Darier levels a scathing critique of what he calls the "naturalistic" environmental ethic that scholars like Rolston (2002) and J. Baird Callicott (1980, 1989) propose, on the grounds that naturalistic ethics are just more "normative yardstick[s] to impose … on human behaviour and values" (1999a, p. 217). Darier's issues with naturalistic ethics, grounded in Foucault's (1978, 2003) critiques of universalizing norms, center on how universal claims that "transcend the cultural and the historical" can be leveraged against humans and the natural world (Darier, 1999a, pp. 217–218). In both the *History of Sexuality Volume 1: An Introduction* (1978) and *Society Must be Defended* (2003), Foucault clearly articulates his criticisms of normalizing practices entrenched in biology, science, or medicine through his discussion of "biopolitics" that grants the state the power over life through "making live":

> For the first time in history, no doubt, biological existence was reflected in political existence; the fact of living was no longer an inaccessible substrate that only emerged from time to time … part of it passed into knowledge's field of control and power's sphere of intervention. (Foucault, 1978, p. 142)

Foucault establishes these key connections among the biological, the political, and the power of the state to intervene that serve as the foundation for Darier's criticisms of naturalistic ethics. For Darier, naturalistic ethics are problematic for their potential to broaden biopolitics to "all life forms" through "ecopolitics," or an "attempt to extend control ('management') to the entire planet" (Darier, 1999b, p. 23).

Darier's (1999a) response, a "green aesthetics of existence," is entrenched in Foucault's aesthetics of self. According to Darier, Foucault describes "different aspects of ethics as understood since the Greeks" as: (1) the morality in which individuals are "normalized"; (2) the morality of behaviors in relation to recommended rules or values; and (3) the "ethics of 'conduct'" or reflexive work on the self (1999a, p. 227). He goes on to argue that a Foucauldian environmental ethic would focus on the *human* self through practices of "'self-reflection, self-knowledge, self-examination' of the existing limits of what constitutes the 'environment' and the individual's conduct *vis-à-vis* the 'environment' and *vis-à-vis* oneself" (1999a, 227).[9] Darier (1999a) stresses that the objective of his green aesthetics of existence might not actually be to "save the planet," but that processes of self-reflection might lead to valuing of the environment or natural world. His focus, then, involves a shift from what he views as a potentially totalizing ecopolitical ethic grounded in naturalism that might authorize the state to intervene in the lives of humans and nonhumans to an ethics of the individual self. As example at the Aquarium, rather than focusing on the biological aestheticization of sharks that Darier (1999a) would view as a Foucauldian ecopolitics (akin to biopolitics) for its management of life through the normalizing of the "beautiful" and the "natural" and connecting those qualities to value, a green aesthetics of self would focus on work to aestheticize the *human* individual through self-reflection.

While Darier's approach is laudable for his focus on avoiding the potential pitfalls of strictly naturalistic ethics (i.e., possibly ecopolitical ethics qua Foucault's biopolitics), it nonetheless is flawed due its humanism. As Mick Smith (2001) points out, Darier's green aesthetics of self conceptualizes the environment as incidental to the work of "self-fashioning" making it anthropocentrically reductive (2001, p. 58). Smith (2001), following Foucault's work (1992) in the *The History of Sexuality Volume 2: The Use of Pleasure*, argues that strictly focusing on the moral actions of the self and not on those actions forged through a "relationship with the reality in which it [the self] is carried out" (Foucault, 1992, p. 28) ignores potential "environmental dimensions" in Foucault's work (Smith, 2001, p. 59). That is, Smith (2001) contends that a Foucauldian environmental ethic would not be limited to a green aesthetics of self (Darier, 1999a); rather, it must also consider the "reality" in which the self lives, that is, the natural world. He places importance on the recognition of an Other (i.e., someone or something in the natural world) "more than and different from me," which is a self-fashioning *in relation* to the natural world (2001, p. 61). This approach distinguishes his ethic from Darier's (1999a, 1999b) in that the natural world is not incidental to self-fashioning; it also digresses from naturalistic ethics because of its focus on the human self (Smith, 2001). Resonating with the last several decades of feminist, posthumanist, and interspecies scholarship in environmental education (Bell & Russell, 2000; Fawcett, 2013; Nxumalo & Pacini-Ketchabaw, 2017; Oakley et al., 2010; Pacini-Ketchabaw & Nxumalo, 2015; Pacini-Ketchabaw & Taylor, 2015; Pedersen, 2004, 2010a, 2010b, 2013; Russell & Bell, 1996; Spannring, 2017; Taylor & Pacini-Ketchabaw, 2015, 2016), such a position recognizes that the human self becomes through relations with other living entities—and that the content and quality of those relations matter.

Despite their differing philosophical and ontological commitments, Rolston (2002), Smith (2001), and Yusoff (2011) agree in their cautioning against an anthropocentric aestheticized Other fit for human consumption—like the Aquarium's disembodied shark. Smith's Other, for example, formed through ethical relations resisting desire and possession of the nonhuman, is instead "more than and different from me" (2001, p. 61). The Aquarium's anthropocentric and naturalistic discourses teeter on "normalizing" biopolitical practices, reducing sharks to aspects of their biology capable of aestheticization and consumption (Darier, 1999a, 1999b; Foucault, 2003). That is, the hidden curriculum of the Aquarium's pedagogical practices actually constructs a disembodied shark aesthetically and biologically normalized to human standards of beauty and fitness in order to evoke care and concern for real sharks. However, if value is derived in this manner, then what becomes of those beings who fall short of the aesthetic norm? Rolston (2002) addresses this issue to some degree in arguing that aesthetics can be derived not only through artistic beauty, but also wonder in evolutionary achievements. Following Darier (1999a, 1999b) and his interpretation of Foucault, however, grounding any ethic in the "universal truths" of science (i.e., evolution) is risky due to the potential of biopolitical or ecopolitical management. That is, Darier (1999a, 1999b) would argue that if the value of organisms (including humans) is derived from biologically-based aesthetic criteria, then the state (or other entities vested with formalized power) might be granted authority to intervene and manage bodies and populations in order to "make live" (Foucault, 1978, 2003) in very specific ways, which now include neoliberal ones intimately tied to economics (Fletcher, 2010).

It is essential to highlight here, that "making live" at the Aquarium paradoxically involves a Foucauldian "death" for the Aquarium's small sharks. While Foucault (1978, 2003) did not address speciesism, or "discrimination against nonhumans" (Midgley, 1983, p. 65), he advanced clear arguments about the role of racism in deciding "what must live and what must die" (2003, pp. 255–257). In his view racism functions to fragment humans, based on biological normalizing categories, into those who can live (the biologically superior) and those who must die (the biologically inferior "Other"), resonating with the practices of eugenics (Subramaniam, 2014). Foucault's "murderous functions" of the state or other institutions vested with state-like power do not simply involve killing, but other forms of death such as "exposing someone to death, increasing the risk of death for some people, or, quite simply, political death, expulsion, rejection, and so on" (Foucault, 2003, p. 256). Applying Foucault's biopolitical conception of racism to species, recognizing that racism against human beings is not the same as, or equivalent to speciesism, I argue that the Aquarium's admirable attempts to invoke care and concern for the disembodied shark actually stem from anthropocentric and speciesist discourses that justify biopolitical or ecopolitical management of shark bodies and ultimately lead to a Foucauldian "death" for those sharks residing at the Aquarium.[10]

Staff at the Aquarium draw on aestheticizing discourses grounded in biology in order to ask visitors to feel wonder, compassion, and care for Aquarium sharks. As stand-ins for all sharks, portrayed in the popular media as savage and brutal (and therefore biologically inferior Others), the small sharks are re-fashioned as the disembodied shark—Others fit for human consumption. Paradoxically, while the small sharks' "evolutionary achievements" are corporeally and materially present in the shallow pools at the Aquarium, they bear little meaning to the sharks themselves in an environment where hunting, mating, caring for young, and other behaviors have been expunged; instead, "fitness" is intricately connected with appealing to human sensibilities. This shark is not an Other fashioned as "more than and different from me," as Smith (2001) would advocate. Instead, the small sharks are discursively "expelled" from their own species, families, and orders, as their meaning and value are derived from the way their "evolutionary achievements" provide pleasurable experiences for people, and not for the ways in which they provide the actual sharks with skills for a good shark life, whatever that might entail. In a Foucauldian (2003) "political death," resulting from the Aquarium's conservation efforts, these sharks are set aside as biopolitical representatives and conservation martyrs (e.g., Lloro-Bidart & Russell, 2017) of the disembodied shark they save through their own regulation, expulsion, and aesthetic appeal to humans. That is, in order to make the biologically superior disembodied shark live, the inferior small sharks must be subject to political death. The biopolitical management Darier (1999a, 1999b) cautions against, therefore, manifests itself at the Aquarium as the capture and regulation of small shark bodies is paradoxically justified in what appear as discourses of compassion and care in a conservation education project. In the final chapter of this book, I draw out the implications of these practices specifically for the fields of environmental education and zoo and aquarium interpretation.

Notes

1. Until 2011, paid Interpreters staffed the Shark Lagoon. As a cost-saving measure, Aquarium management decided to hand the exhibit over to volunteers to reduce the number of paid staff needed each season.
2. The notion of "ecosystem stability" is fraught with criticism, however. Hettinger and Throop (1999) argue, for example, that there is substantial scientific evidence documenting ecosystem instability. They suggest valuing "wildness," which could be contextually defined. "Ecosystem services" is also a problematic term, insofar as it neoliberally suggests that ecosystems and their individual members provide services that can be commodified and marketized, only giving them economic value.

3. Sharks do not typically consume entire human bodies, although their bites can be fatal. Quammen (2003) chronicles the lives of people living with crocodiles, known to gobble up intact bodies of humans and other large animals.
4. While there are documented shark attacks, it is likely that the "man-eating" status of sharks in popular film has been exaggerated.
5. I summarize the diver's speech from field notes because I had consent to audio-record and utilize the education staff member's exact words, but not the permission of the diver. See Teresa Lloro-Bidart (2014) for a full explanation of methods and methodology.
6. In this chapter, I focus on the discourses of staff members who utilize such aestheticizing practices; not all staff do this because of the un-scripted nature of "on-mic" discourses at the Aquarium.
7. Recall that none of the sharks at the Aquarium hunt for food, which means that visitors never have to view sharks ripping other animals to pieces. In theory, the sharks *could hunt*, as they inhabit both Blue Cavern and Big Trop with fish they might consume in the wild. These sharks do not hunt, however, because staff feed them well. Small (2012) points out that being a powerful predator is typically an attractive trait; the attractiveness diminishes, however, when the animal is viewed as predator of humans.
8. The possibility of emotional human-animal relationships is also traditionally reserved for mammals and birds in Western conservation discourses. Bekoff (2007), for example, describes the importance of animal emotions to human health and happiness. His cases, however, all involve mammals. One notable exception is Porcher's *My Sunset Rendezvous, Crisis in Tahiti*, where the author, a wildlife artist turned conservationist, describes the social relationships she developed with sharks in a seven-year ethological study in Tahiti.
9. Note that Foucault did not develop an environmental ethic in any of his work. Darier's (1999a) task, then, is mapping out what a Foucauldian environmental ethic *might* look like, as others have done (see Huffer (2015) for a feminist environmental ethic based on Foucault's work).
10. For more on connections between racism and speciesism see Deckha (2010, 2012), Kim (2015), and Ko and Ko (2017).

References

Appleby, R., & Pennycock, A. (2017). Swimming with sharks, ecological feminism, and posthuman language politics. *Critical Inquiry in Language Studies*. Advance online publication. doi 10.1080/15427587.2017.1279545.

Bauernfeind, E. (2017, April 21). Piles and piles of sharks. *New England Aquarium*. Retrieved from http://www.neaq.org/blog/piles-piles-sharks/. Accessed September 7, 2020.

Bekoff, M. (2007). *The emotional lives of animals*. Novato, CA: New World Library.

Bell, A., & Russell, C. (2000). Beyond human, beyond words: Anthropocentrism, critical pedagogy, and the poststructuralist turn. *Canadian Journal of Education, 25*(3), 188–203.

Boileau, E., & Russell, C. (2018). Insect and human flourishing in early childhood education: Learning and crawling together. In A. Cutter-Mackenzie, K. Malone, & E. Barratt Hacking (Eds.), *Research handbook on childhoodnature: Assemblages of childhood and nature* (pp. 1323–1338). New York, NY: Springer.

Callicott, J. B. (1980). Animal liberation: A triangular affair. *Environmental Ethics, 2*, 311–338.

Callicott, J. B. (1989). *In defense of the land ethic*. Albany, NY: State University of New York Press.

Darier, E. (1999a). Foucault against environmental ethics. In E. Darier (Ed.). *Discourses of the environment* (pp. 217–240). Malden, MA: Blackwell Publishers.

Darier, E. (1999b). Foucault and the environment: An introduction. In E. Darier (Ed.). *Discourses of the environment* (pp. 1–33). Malden, MA: Blackwell Publishers.

Deckha, M. (2010). The subhuman as an agent of violence. *Journal of Critical Animal Studies, 8*(3), 28–51.

Deckha, M. (2012). Toward a postcolonial, posthumanist feminist theory: Centralizing race and culture in feminist work on nonhuman animals. *Hypatia, 27*(3), 527–545.

Dobson, J. (2011). Fun, fascination and fear: Exploring the construction and consumption of aquarium shark exhibits. In W. Frost (Ed.), *Zoos and tourism: Conservation, education, entertainment?* (pp. 85–99). London, UK: Channel View Publications.

Douglas, L. R., & Veríssimo, D. (2013). Flagships or battleships: Deconstructing the relationship between conflict and conservation flagship species, *Environment & Society: Advances in Research, 4*, 98–116.

Fawcett, L. (2013). Three degrees of separation: Accounting for naturecultures in environmental education research. In R. B. Stevenson, M. Brody, J. Dillon, & A. E. J. Wals (Eds.), *The international handbook of research on environmental education* (pp. 409–417). New York, NY: Routledge.

Fletcher, R. (2010). Neoliberal environmentality: Towards a postructuralist political ecology of the conservation debate. *Conservation and Society, 8*(3), 171–181.

Foucault, M. (1978). *The history of sexuality volume 1: An introduction*. New York, NY: Vintage Books.

Foucault, M. (1992). *The history of sexuality volume 2: The use of pleasure*. Harmondsworth, UK: Penguin.

Foucault, M. (2003). *Society must be defended*. New York, NY: Picador.

Foucault, M. (2010). *The Foucault reader*. New York, NY: Vintage Books.

Haraway, D. J. (2008). *When species meet*. Minneapolis, MN: University of Minnesota Press.

Hettinger, N., & Throop, B. (1999). Refocusing ecocentrism: De-emphasizing stability and defending wildness. *Environmental Ethics, 21*(1), 3–21.

Huffer, L. (2015). Foucault's fossils: Life itself and the return to nature in feminist philosophy. *Foucault Studies, 20*, 122–141.

Kim, C. J. (2015). *Dangerous crossings: Race, species, and nature in a multicultural age.* Cambridge, MA: Cambridge University Press.

Ko, A., & Ko, S. (2017). *Aphro-ism: Essays on pop culture, feminism, and black veganism from two sisters.* New York, NY: Lantern Books.

Lemelin, R. H., Harper, R. W., Dampier, J., Bowles, R., & Balika, D. (2016). Humans, insects, and their interaction: A multi-faceted analysis. *Animal Studies Journal, 5*(1), article 5.

Leopold, A. (1949). *A Sand County almanac: And sketches here and there.* Oxford, UK: Oxford University Press.

Lloro-Bidart, T. (2014). *Reassembling the "environment": Science, affect, and multispecies educative practice at the Aquarium of the Pacific.* (Unpublished doctoral dissertation). Riverside, Riverside, California: University of California.

Lloro-Bidart, T. (2018). Cultivating affects: A feminist posthumanist analysis of invertebrate and human performativity in an urban community garden. *Emotion, Space, and Society, 27*, 23–30.

Lloro-Bidart, T. & Russell, C. (2017). Learning science in aquariums and on whale-watching boats: The political deployment of other animals. In M. P. Mueller, D. J. Tippins, & A.J. Stewart (Eds.), *Animals in science education: Ethics, curriculum, and pedagogy* (pp. 41–50). New York: Springer.

Lukasik, J. M. (2013). Embracing my escape from the zoo: A (sometimes) true account of my curricular inquiry. *Green Theory & Praxis Journal, 7*(1), 3–16.

Luke, T. W. (2000a). Beyond birds: Biopower and birdwatching in the world of Audubon. *Capitalism Nature Socialism. 11*(3), 7–37.

Luke, T. W. (2000b). The Missouri Botanical Garden: Reworking biopower and flora-power. *Organization* Environment, *13*(3), 305–321.

Malamud, R. (2017). The problem with zoos. In L. Kalof (Ed.), *The Oxford handbook of animal studies* (pp. 397–410). New York, NY: Oxford University Press.

Marietta, D. E. (1995). *For people and the planet.* Philadelphia, PA: Temple University Press.

Marino, L, Lilienfeld, S. O., Malamud, R., Nobis, N., & Broglio, R. (2010). Do zoos and aquariums promote attitude change in visitors? A critical evaluation of the American Zoo and Aquarium Study. *Society & Animals, 18*(2), 126–138.

Midgley, M. (1983). *Animals and why they matter.* Athens, GA: The University of Georgia Press.

Myers, R. A., Baum, J. K., Shepherd, T. D., Powers, S. P., & Peterson, C. H. (2007). Cascading effects of the loss of apex predatory sharks from a coastal ocean. *Science, 315*(5820), 1846–1850.

Nxumalo, F., & Pacini-Ketchabaw, V. (2017). "Staying with the trouble" in child-insect-educator common worlds. *Environmental Education Research, 23*(10), 1414–1426.

Oakley, J., Watson, G., Russell, C., Cutter-Mackenzie, A., Fawcett, L., Kuhl, G., Russell, J., van der Waal, M., & Warkentin, T. (2010). Animal encounters in environmental education research: Responding to the question of the animal. *Canadian Journal of Environmental Education*, *15*, 86–102.

Pacini-Ketchabaw, V., & Nxumalo, F. (2015). Unruly raccoons and troubled educators: nature/culture divides in a childcare centre. *Environmental Humanities*, *7*, 151–168.

Pacini-Ketchabaw, V., & Taylor, A. (2015). Unsettling pedagogies through common world encounters: Grappling with (post)colonial legacies in Canadian forests and Australian Bushlands. In V. Pacini-Ketchabaw & A. Taylor (Eds.), *Unsettling the colonial places and spaces of early childhood education* (pp. 43–62). New York, NY: Routledge.

Panoch, R., & Pearson, E. (2017). Humans and sharks: Changing public perceptions and overcoming fear to facilitate shark conservation. *Society & Animals*, *25*(1), 57–76.

Pedersen, H. (2004). Schools, speciesism, and hidden curricula: The role of critical pedagogy for humane education futures. *Journal of Futures Studies*, *8*(4), 1–14.

Pedersen, H. (2010a). *Animals in schools*. West Lafayette, IN: Purdue University Press.

Pedersen, H. (2010b). Is 'the posthuman' educable? On the convergence of educational philosophy, animal studies, and posthumanist theory. *Discourse: Studies in the Cultural Politics of Education*, *31*(2), 237–250.

Pedersen, H. (2011). Counting affects: Mo(ve)ments of intensity in critical avian education. *Canadian Journal of Environmental Education*, *16*, 14–28.

Pedersen, H. (2013). Follow the Judas sheep. Materializing post-qualitative methodology in the zooethnographic space. *International Journal of Qualitative Studies in Education*, *26*(6), 717–731.

Quammen, D. (2003). *Monster of god.* New York, NY: W.W. Norton and Company.

Rabinow, P., & Rose, N. (2006). Biopower today. *Biosocieties*, *1*(2), 195–217.

Rolston, H. (2002). From beauty to duty: Aesthetics of nature and environmental ethics. In A. Berleant (Ed.), *Environment and the arts: Perspectives on environmental aesthetics* (pp. 127–141). Hampshire, UK: Aldershot.

Russell, C. (2005). "Whoever does not write is written": The role of "nature" in post-post approaches to environmental education research. *Environmental Education Research*, *11*(4), 433–443.

Russell, C., & Bell, A. (1996). A politicized ethic of care: Environmental education from an ecofeminist perspective. In K. Warren (Ed.), *Women's voices in experiential education* (pp. 172–181). Dubuque, IA: Kendall Hunt.

Small, E. (2012). The new Noah's Ark: Beautiful and useful species only. Part 2. The chosen species. *Biodiversity*, *13*(1), 37–53.

Smith, M. (2001). The face of nature: Environmental ethics and the boundaries of contemporary social theory. *Current Sociology*, *49*(1), 49–65.

Spannring, R. (2017). Animals in environmental education research. *Environmental Education Research*, *23*(1), 63–74.

Subramaniam, B. (2014). *Ghost stories for Darwin: The science of variation and the politics of diversity*. Chicago, IL: The University of Chicago Press.

Taylor, A., & Pacini-Ketchabaw, V. (2015). Learning with children, ants, and worms in the Anthropocene: Toward a common world pedagogy of multispecies vulnerability. *Pedagogy, Culture, and Society, 23*(4), 507–529.

Taylor, A., & Pacini-Ketchabaw, V. (2016). Kids, raccoons, and roos: Awkward encounters and mixed affects. *Children's Geographies 15*(2), 131–145.

Yusoff, K. (2011). Aesthetics of loss: Biodiversity, banal violence, and biotic subjects. *Transactions of the Institute of British Geographers, 37*(4), 578–592.

Chapter Six

Affective Labor in the Lorikeet Forest

Introduction

Although a number of environmental education scholars have employed feminist methodologies and theories during the last two decades to empirically study teaching and learning, the field generally underutilizes feminist perspectives (Gough, 2013; Gough, Russell, & Whitehouse, 2017, 2018; Gough & Whitehouse, 2018), particularly as tools to understand human-animal relationships in the zoo and aquarium context (for exceptions see Lloro-Bidart, 2017a, 2018a; Russell, 2000; Warkentin, 2010, 2012). Further, as highlighted in prior chapters, the field has yet to meaningfully engage with political ecology, which analyzes human-nature relations/relationships as situated within various political contexts and complex networks (Lloro-Bidart, 2015, 2017a, 2017b; 2018a; Meek, 2015a, 2015b, 2015c; Meek & Lloro-Bidart, 2017), with power as the primary mediating construct (Biersack, 2006). And finally, while scholars are increasingly concerned with the affective and emotive dimensions of environmental education, there is much that remains unknown (e.g. Russell & Oakley, 2016). This chapter, therefore, brings a feminist posthumanist political ecology to bear on the Aquarium's Lorikeet Forest (The Forest) Exhibit and the Diving Birds Exhibit in order to explore what feminist posthumanist political ecology offers animal-focused education, particularly in zoos and aquariums.

In environmental education, some work has demonstrated the value of feminist postructuralism in critically challenging who can be a subject of inquiry and an agent of knowledge (e.g., Barrett, 2005; Bell & Russell, 2000; Fawcett, 2013; Gough, 2013; 2018; McKenzie, 2005; Russell, 2005). However, feminist postructuralism tends to focus exclusively on the discursive practices of humans, which, as Constance Russell points out, ignores other beings who are "busily getting on with their own existence, yet ... are profoundly, materially, impacted by our inscriptions" (2005, p. 435). Russell shows how the "co-construction," of nonhuman and human voices is often "denied ... not sought, and is not even heard" (2005, p. 435). In this chapter I do not claim that a feminist posthumanist political ecology allows one to discursively converse with nature or other animals, but I do suggest that in drawing on posthuman and multispecies perspectives that tend to the political and ethical contexts of knowledge making, a feminist posthumanist political ecology does the following: (1) considers a multispecies rather than an exclusively human community of knowers (Fawcett, 2005, p. 276); (2) gives epistemological, theoretical, and methodological attention to nonhumans politically deployed in educational processes (Lloro-Bidart, 2018b; Lloro-Bidart & Russell, 2017; Pacini-Ketchabaw, Taylor, & Blaise, 2016; Ogden, Hill, & Tanita, 2013; Russell & Bell, 1996); and (3) focuses on relational ways of knowing and nonhuman action and lived experience (Lorimer & Srinivasan, 2013).[1] This perspective engages feminist postructuralism's attention to relational ways of knowing, but focuses more explicitly on how nonhuman animals are subjects of inquiry and agents of knowledge as they participate in learning experiences and are materially (and in some cases emotionally) affected by them.

I begin with a brief overview of the exhibits germane to this chapter, situating the current case of Aquarium volunteer staff member, Jennie, and the Aquarium's diving birds and lorikeets within this greater institutional frame. Next I demonstrate how, despite working in a political context that fashions birds and other animals as objects for display and/or interaction, Jennie comes to see them as subjects, or "persons" because of her direct experiences caring for them. I conclude by outlining the major tenets of this feminist posthumanist political ecology, which include attunement to: (1) the lived experiences of animals and humans, including human/animal agency, human discursive practices, and the role of the material environment; (2) embodied, emotional, and affective ways of knowing that allow humans to see nonhumans as persons; and (3) the political ecological and political economic contexts in which these human-animal relationships occur (e.g., Lloro-Bidart, 2017a; 2018b).[2] The purpose of this framework is to contribute to a growing conversation in environmental education regarding human-nonhuman relationships and to explore the utility of feminist posthumanist political ecologies in animal-focused education practice.

Jennie and the Aquarium's Birds

In this chapter, I focus primarily on a case study of one volunteer staff member, Jennie, and her complex relationships with the Aquarium's captive rainbow lorikeet and puffin populations, although I also situate this case within the larger institutional context. Recall that Jennie was also featured in Chapter Four when I analyzed interpretation in the Forest Exhibit. Although the "target" of Aquarium educational programming is Aquarium visitors, Aquarium staff members also learn in this context. Due to the transient nature of Aquarium visitorship, it would not be possible for me to collect such rich ethnographic data on visitor experiences but I was able to do so with staff experiences. Therefore, this chapter seeks to capture, through the portrait of one staff member, the potentialities of human staff relationships at facilities like the Aquarium. Below, I detail the epistemological and methodological reasons for focusing on Jennie and the birds she cares for at the Aquarium.

Aquarium lorikeets live in one of three places: the Forest Exhibit (a 3,200 square foot outdoor aviary), the Hut (a small off-exhibit structure inside the Forest Exhibit), or the Barn (a large off-exhibit building housing lorikeets and other program animals).[3] The number of lorikeets on-exhibit fluctuates widely as nesting, ill, injured, young, old, or "difficult" birds are shuffled amongst the various spaces. The Aquarium's tufted and horned puffins reside in the Northern Pacific Gallery, an indoor exhibit hall designed to mimic life in the Bering Sea. Together with two other species of Alcids (birds of the auk family), they comprise the animals of the Diving Birds Exhibit. According to Aquarium signage, the original tufted puffins came from other zoos and aquariums and from the Lawrence Island in the Bering Sea. The horned puffins came from Alaska. Both original populations, then, were captured from the wild to serve edutainment ends at the Aquarium.

When I began my study, Jennie had volunteer roles in both the Husbandry and Education Departments, but toward the end she decided to focus primarily on her Husbandry work, especially with the lorikeets. She explained that once she was able to secure her first Husbandry position, her priority became spending time with and caring for the Aquarium's birds. For Jennie, Education involved too much interacting with and/or entertaining visitors and not enough attention to the lives of the captive animals. By the time my study concluded, Jennie volunteered two to three full days per week caring for Aquarium animals.[4]

I highlight Jennie's experiences with the lorikeets and puffins for several reasons. First, Jennie acquired her first Husbandry position at the beginning of my fieldwork and just two months prior to our first interview. This afforded me the opportunity to follow Jennie throughout the first year of what would become several volunteer Husbandry positions. Second, while a few volunteer and paid staff members held positions in Husbandry and Education, Jennie more openly and explicitly identified with her Husbandry work. Further, Jennie eagerly and openly shared her thoughts with me about the Aquarium, and more specifically about the birds (primarily lorikeets) with whom she spent vast amounts of time. Our conversations were honest—sometimes raw and emotional—and it was obvious that Jennie's work mattered to her beyond what one might consider a regular "job" or volunteer position. I often shadowed her because of her willingness to be candid—so much so that she one day joked, "That's why you follow me around!" Jennie also afforded me access to behind-the-scenes Aquarium spaces I otherwise would not have viewed. Through these experiences, I saw the daily and often dirty work Jennie willingly performs, such as scrubbing feces off cages, preparing food, and hosing down enclosures, through her own eyes; I also glimpsed what life is like for off-exhibit Aquarium animals. And finally, while many other staff members certainly describe caring about animals, the oceans, or the environment more generally, the personal connection Jennie developed with the diving birds and lorikeets was, as I describe in later sections, unique among staff who worked in both the Education and Husbandry departments at that time. I therefore chose to thickly describe (Geertz, 1973) Jennie's case and use it to build theory to draw out wider implications for the field of environmental education, particularly feminist and posthumanist approaches to teaching about and studying human-animal relationships.

Laboring Affectively: Seeing Birds as Persons

In this section, I present ethnographic data that demonstrates the complex ways in which Jennie's direct experiences caring for Aquarium birds are connected to her viewing them as persons. These interview excerpts generally reflect all of my interactions with Jennie, whether we chatted in the Aquarium staff break room, in the space of the Forest, or as I followed her from station to station when she volunteered in the Education Department.[5] To address Laura Bear, Karen Ho, Anna Tsing, and Silvia Yanagisako's (2015) critiques of affective labor perspectives, I remain attuned not only to the affects Jennie's work produces, but also the materiality of these relations and "products" as I refer to her affective labor.[6] In the final section, I use these data to describe a feminist posthumanist political ecology as I consider: (1) the lived experiences of animals and humans, as "networks of learners" (Rocheleau, 2015, p. 57) and "communities of knowers" (Fawcett, 2005, p. 76); (2) embodied and emotional ways of knowing (Donovan, 2006) that allow humans to see nonhuman animals as persons (Milton 2002,

2005; Smith 2001); and (3) the political-economic and political-ecological contexts in which these human-bird relationships occur (Donovan, 2006; Fawcett, 2005).

In Jennie's Husbandry position working with Aquarium diving birds, which she describes in the interview excerpt below, she goes on at length about Tuxedo, a horned puffin.[7] Notice how this passage captures Tuxedo's contributions to this multispecies community of knowers—he learns how to be with humans as much as they learn how to be with him:

> *My first day in Husbandry, we have this puffin and his name is Tuxedo [Tuxedo does not have a mate and is lonely and aggressive]. He thinks that we're birds or he's human, but he wants [to be with] Joe [paid Husbandry staff member; Jennie's boss]... He gets up in Joe's lap and he cuddles Joe ... He [Joe] opens the thing and Tuxedo hops out of the cage and he's walking around the floor ... And I'm standing there and I'm not afraid of birds and I'm not going to show any fear. You don't do that with birds. He waddles right around the back of me and he goes chomp [makes a chomp sound] and he bit me and drew blood. He bit me right on the heel right above my sock. And I'm like, "Oh!..." I said, "Okay, I'm gonna take that as my initiation. Tuxedo, you and I are going to be buds. It might take a few years, but you're not gonna walk up behind me like that again." And so I take care of him every week and we just put him in with the other birds and I've been watching him, too, to make sure he's not being aggressive. But that was my first day in Husbandry. Tuxedo and I have a special connection. He comes out; he wants out; he wants to play. You know, just the rules all belong to him.*

As this vignette demonstrates, in just two short months working in Husbandry Jennie describes taking care of and having a special connection with Tuxedo despite having to handle aggressive behavior. She quickly learns that "the rules all belong to him"—a political statement given that in the captive environment in which Tuxedo resides, most of the formalized, institutional rules that govern Tuxedo's existence are in fact not his own.

At the national level, the Association of Zoos and Aquariums (AZA) and the Animal Welfare Act guide Tuxedo's care.[8] Institutionally, administrators and managers make final decisions about which animals remain at the Aquarium despite aggressive or unmanageable behavior, including resistance to training; they decide which animals get the opportunity to breed and with whom, or even if the animals get to have platonic relationships with other members of their own kind; and finally, they also decide if visitors can tactually interact or only gaze upon specific animals.[9] That is, at the macro-institutional level administrators and managers decide which animals are politically deployed in a variety of endeavors, including educational ones, and to a certain extent how these animals qualitatively figure into these processes. Together with the larger national bodies governing zoos and aquariums, they purposefully do not challenge animal captivity and instead work to ensure that visitors encounter ordered spaces where it is possible to perceive that the animals involved in edutainment processes are well cared for and enjoying a vital life (Braverman, 2013; Chrulew, 2011; Hanson, 2002). While there is fierce debate regarding whether such a captive life constitutes flourishing, simply existing, or suffering (Chrulew 2011; Rudy, 2011) as discussed in Chapters Four and Five, Jennie's acknowledgement that Tuxedo makes all the rules demonstrates that at the ground level, nonhuman animals participate in learning processes in such a way that staff who care for them *can* view them as active agents, or as persons. Jennie's lived experiences caring for Tuxedo foster emotional connections to him in spite of or because of his rule-making, and emerge as critical to these understandings.

Several months after Jennie began her volunteer work with the diving birds, she acquired a Husbandry position with the Aquarium's captive lorikeet population—a position she adored despite being conflicted about the kinds of relations visitors had with the lorikeets. Recall that the institution sells cups of nectar to visitors for $3 in order to generate revenue and foster visitor-lorikeet interaction. Many visitors, then, enter the Forest expecting to feed birds in what amounts to a highly commodified and biopolitically regulated encounter, as discussed in Chapter Four. This situation results in several issues that concern Jennie (and also concern other staff members and me). Sometimes visitors chase the birds with their cups of nectar in an attempt to force interaction; other visitors poke or prod them; and still others unwittingly pass into nesting spaces as they traverse the Forest, resulting in situations where many Aquarium staff feel conflicted about how to keep the lorikeets safe while maintaining customer satisfaction. In addition, the feeding of the nectar itself disrupts bird territorial patterns. According to Jennie's novice ethological knowledge, hungry birds sometimes pass into other birds' defined spaces in the Forest, resulting in fighting or disagreements.

These visitor-lorikeet interactions contrast markedly with the labor Jennie performs for/with the lorikeets and diving birds. For example, one day toward the end of my ethnographic data collection, Jennie volunteered to take me on an informal, behind-the-scenes tour of the "Barn," a place I had not previously visited. During this tour, she bustled about the facility in a manner very similar to the many times I had witnessed her cleaning and working in the space of the Forest. She pointed out some of her cleaning and feeding duties, told me stories about various lorikeets in the facility, and also went on to explain why certain birds had particular housing arrangements—that is, some had paired up and were mating or caring for eggs, some get along poorly with other birds, and some had acquired illnesses and were being cared for by staff.

The same day Jennie took me on the tour of the Barn, I conducted a follow-up interview with her. At that time, she had just over a year experience in Husbandry and was also working in several different departments. I hoped, therefore, to get a sense of how she felt about her work in Education versus her work in Husbandry, after having substantial experiences in both roles. Drawing on Walter Doyle's (1992) research in formal educational spaces regarding the differences between the institutional and enacted curriculum, I was also interested in how Jennie perceived her work in relation to institutional aims.[10] Notice how she describes Feathers, a juvenile lorikeet she came to know well through serving as her caretaker for the past six months. Five months prior to this interview, during an impromptu behind-the-scenes tour inside the Hut, I met Feathers; at that time she was patchily covered in grayish-white down feathers and the green and blue feathers of lorikeet adulthood:

Teresa: Do you think that any of the messages you hope to convey are different from that of the institution?

Jennie: I don't know that my messages, that the messages that I would like to see us convey, are different. Honestly, I feel like, in general, there's the public face of the

> *Aquarium and then there's real life. And they're not necessarily the same thing. Being in Husbandry you see more of that. It's a little bit more [of] reality. And so you get much more day-to-day and you get to know the animals and you have, you develop relationships with them. And it's not the same as ... I don't focus on the visitors. And there are a lot of challenges that, I feel with the animals, in general, the Aquarium's perspective and their goals and their missions and all that are great. Sometimes I think there's a gap between what they say and what they do. It's harder in Husbandry because you kind of see the realities.*

A bit later in the interview, she illustrates the depth of her coming to know the birds:

> *Jennie: We have all these little babies. Feathers, you remember Feathers?*
>
> *Teresa: I remember Feathers.*
>
> *Jennie: Feathers, Feathers had—Feathers is still the cutest bird in there, but Feathers now is all cuddly ... and Feathers comes up and Feathers will play and she gets your hand and you can hear Feathers over all the different birds, she has a different—it's a she by the way. I was sure she was a boy, but she's a girl. And she has this little melodic voice that you can hear very differently; and it's just so cute.*

While Jennie opens this passage by explaining that she does not believe her own aims and those of the institution diverge much, she goes on to concretely describe how the institution focuses on "the visitors" while she labors for and simultaneously develops emotional and affective relationships with the birds. Through her "day-to-day" work in Husbandry, she has come to know the lorikeets as persons and has developed relationships with them, much like her "special connection" with Tuxedo. While laboring to generate a material product of sorts, that is, healthy animals fit for display and interaction with visitors at a neoliberal edutainment institution (Lloro-Bidart, 2017a, 2017b), Jennie also affectively and emotionally engages with the lorikeets in a sort of relational epistemology that undoes the subject-object divide pervading educational spaces wherein only humans are considered legitimate knowers (Bell & Russell, 2000; Oakley et al., 2010; Pedersen, 2004, 2010a, 2010b). In these relationships she comes to know and see the animals as persons, both because she recognizes the positive and negative characteristics of personhood in them (e.g., playing, singing, physically interacting, biting, acting aggressively, making rules) and because of the affective labor she performs in order to care for them. That is, Jennie does not simply make sense of her relationships with the birds, but she is actually affected by them.[11] While the macro-political context in which these interactions occur indeed constrains Jennie and the birds (the birds far more than Jennie), the preceding analysis demonstrates that there are products, or affects, that constitute biopower from below (Hardt, 1999; Singh, 2013) as Jennie develops relationships with them and cares for them as persons.

Toward a Feminist Posthumanist Political Ecology of Education

In the preceding analysis I considered: (1) the lived experiences of Jennie and the birds she cares for; (2) the embodied and emotional labor Jennie performs that has her come to see the birds as persons; and (3) the political-economic and political-ecological contexts in which these human-animal relationships occurred. Though Jennie's official Husbandry volunteer position requires her to care for Aquarium animals so that the institution remains compliant with the Animal Welfare Act, retains its accreditation with the AZA, and provides edutainment services to the greater Los Angeles area, the current political ecological analysis demonstrates that Jennie produces more than material bodies fit for display and interaction. As she labors for and with Aquarium birds like Tuxedo and Feathers, Jennie develops relationships with them that, while still in neoliberal capitalism's grasp since these relationships are part of producing bodies for display and interaction, also bring about "networks of learners" and "communities of knowers" (for example, Jennie, Tuxedo, and Feathers) who participate in complex, fluid, and emotional relationships.

By the time my research at the Aquarium concluded, the ways in which Jennie's relationships with the birds contributed to her personal and ethical development became quite clear. She specifically cited quitting her work in the Education Department because she viewed it as focusing primarily on the visitor experience, sometimes at the expense of the lives of the animals she had now come to care deeply about. This kind of personal and ethical development, evolving out of sustained, emotional relationships with other animals, is precisely the kind of relating ecofeminist thinkers of the late 20th century advanced and advocated in the context of environmental education (Bell & Russell, 2000; Fawcett, 2002; Russell & Bell, 1996). While my data suggest Jennie's notion of personhood indeed involves a fair amount of what could be characterized as anthropocentrism (e.g., seeing the birds as persons "like me") (Milton 2002, 2005), the data also intimate that she does not view the birds simply as objects and is, in the tradition of feminist political ecology, emotionally connected to them through her embodied experiences with them.

Further, Jennie's admission that she was not going to let Tuxedo "see her fear" because, "You don't do that with birds," indicates that on some level, Jennie recognizes that part of her human-nonhuman ethical relation with Tuxedo (and by extension, possibly other birds if Donovan (2006) is right about her principle of homology) is a posthuman acknowledgement that her understanding of Tuxedo is limited and contained by her own humanness. Such an acknowledgement also intimates that Jennie's novice ethological knowledge—other animals can sense and understand our fear through shared physiological signs, like dilated pupils—plays a role in the complex ways she comes to understand Tuxedo as a person "like her," whom she can partially access through that knowledge, her own emotions and embodiment, and Tuxedo's emotions and embodiment. Yet, resonating with Mick Smith's (2011) and Joanna Latimer's (2013) posthuman partial understandings of the Other, Jennie also clearly understands that there are aspects of his "birdness" (e.g., the "rules" that "belong to him") that she cannot fully know; these inaccessible aspects of his identity also allow her to see him—and possibly other birds—as persons.

The degree to which this framework allows one to access how animals benefit from such relations and relationships is admittedly less clear. AZA member institutions like the Aquarium rather tightly regulate animal care. Therefore, it is plausible that the material benefits Jennie provides (e.g., water, food, clean housing, etc.) do not differ substantively from the care any other human might provide to animals in captivity. But to suggest that the emotional connection Jennie forged with Tuxedo, Feathers, and other Aquarium animals is entirely one-sided would anthropocentrically assume human superiority in cognitive and emotive capacities by ignoring the vast literature in ethology and animal cognition indicating that animals *do* have emotional lives (Balcombe, 2007; Bekoff, 2007; Bradshaw, 2010; Donovan, 2006; Droge, 2012), some of which might be shared in complex ways with humans (see also Rudy, 2011 for an extended discussion of relationships with captive wild animals). Although it may be clearer that such a connection exists between domesticated animals like dogs and humans (Berns, 2012), there is mounting evidence that captive wild animals do experience emotions (both positive and negative) in their interactions with human caretakers (Hosey & Melfi, 2012; Rudy, 2011).

The significance of these embodied and emotional interactions emerge as important in this study, and in other environmental education research involving captive wild animals, particularly when one considers that such wild animals have so many of their "natural" behaviors tightly controlled and regulated and often have limited access to their own species and other species with whom they would interact with in the wild (Braverman, 2013). For example, recall how Jennie described Tuxedo as "lonely" because he did not have a mate, an assertion supported by evidence since puffins tend to breed with the same mate each year. This point is absolutely not intended to valorize human-wild animal relations as a substitute for genuine wild animal-wild animal or species-species relations that exist in the absence of human contact, but rather to argue that the complexities of human-captive wild animal relationships, particularly in spaces that enroll animals in edutainment processes, deserve further theoretical and practical attention.

Kathy Rudy (2011) emphasizes, for example, that with shrinking wild habitat throughout the world, greater attention must be given to how wild animals might live with humans in the future. In her empirical work, she not only explored these kinds of relationships in zoos, but also in roadside sanctuaries and private homes, where some of her informants actually shared their domiciles with animals like tigers. Although I do not entirely agree with Rudy's (2011) assessment that some wild animals are better off living in captive relations with humans, I do concur with her assertion that human relationships with captive wild animals are complex, including in relation to affective ties, and in need of further analytic attention. Thus, while posthuman theories do indeed call us to go beyond the human, such "post" human thinking ought not involve an assumption that the perception of personhood or emotion in other entities constitutes an entirely anthropocentric (or anthropomorphic) endeavor, particularly when studies in ethology and animal cognition increasingly support the idea that we share many similarities with other animals, even those who seem dissimilar, like cephalopods, fish, and birds (Balcombe, 2016; Bekoff, 2007; Bradshaw, 2010; Donovan, 2006). Further, this is also true given that ethologists and posthumanist philosophers have begun to question whether our epistemological and methodological commitments have led us to underestimate the capacity of animals to lead rich cognitive and emotional lives (Despret, 2015; DeWaal, 2016). What posthuman theories accomplish, particularly in concert with feminist political ecological perspectives, is to ask environmental education researchers and practitioners to consider how human and animal lives are complexly *and* politically entangled in such spaces through embodied and affective experiences, echoing Russell and Leesa Fawcett's (2013) recent call in *The International Handbook of Research in Environmental Education* to "move margins" in environmental education by giving voice to those who have been historically absent or at the edges of the field, including other animals.

Conclusions

The framework developed in this chapter, rooted in a relational epistemology, simultaneously tends to the lived experiences of animals and humans, embodied and emotional ways of knowing that allow humans to see nonhuman animals as persons, and the political (economic and ecological) context in which these human-animal relationships occur. This chapter contributes to a growing conversation in environmental education research that considers feminist and posthumanist theories and perspectives from the interdisciplinary field of political ecology. These theories and perspectives uniquely illuminate human-nature relationships in a variety of educational spaces, and address concerns recently expressed by environmental education researchers regarding "incorporating nonhuman animals as stakeholders in our research"

(Oakley et al., 2010, p. 93). While the case presented in this chapter is limited by its focus on one human individual volunteering at one informal learning institution, it foregrounds the ways in which theoretical frameworks in environmental education can be developed through epistemologically, theoretically, and methodologically framing research that counts nonhumans as crucial members of a "community of knowers" (Fawcett, 2005, p. 276) or "network of learners" (Rocheleau, 2015, p. 57). It also calls on environmental education researchers and practitioners to consider how animals become part of our communities and networks, especially through affective relationships. For example, if they become persons, in what ways can we access their personhood and how are we limited? How does political context influence this personhood? What is role of ethological studies and research in animal cognition in our own work? These questions will be explored more fully in the concluding chapter of this book.

Notes

1. In this chapter, I focus primarily on a "multispecies community of knowers" as inclusive of humans and other animals. However, this community also includes insects, plants, fungi, and other living entities.
2. This framework, as I further outline, is deeply indebted to the work of Maneesha Deckha, Josephine Donovan, Kay Milton, Connie Russell, Neera Singh, and Mick Smith.
3. Based on the Aquarium website description, program animals are those which "staff take around the Aquarium to interact with visitors."
4. In order to maintain Jennie's anonymity, I do not fully describe all of her Husbandry positions.
5. Education volunteer staff members generally rotate from exhibit to exhibit every 30 minutes, unless they make special requests to rotate less often.
6. Bear et al. (2015) note that the category of "affective labor" can create a problematic false binary or hierarchy among different forms of labor where industrial labor is viewed as being devoid of affects.
7. All nonhuman animals in this book, like humans, have been assigned pseudonyms to ensure anonymity. To further protect participant identities, these pseudonyms do not necessarily reflect the gender of the persons or animals whom they represent.

8. The AZA, a national accreditation and membership body for zoos and aquariums in the United States, also sets forth standards for management and care of animals in zoos and aquariums. See: https://www.aza.org/uploadedFiles/Accreditation/AZA-Accreditation-Standards.pdf. While the Animal Welfare Act, which "is the only Federal law in the United States that regulates the treatment of animals in research, exhibition, transport, and by dealers" (para 1) governs the lives of bird bodies at the Aquarium, the Act (as a result of complex amendments and provisions since its inception) excludes birds used for laboratory research and those destined to serve as food for humans, such as chickens in confined animal feeding operations (CAFOs) or factory farms. See: https://www.nal.usda.gov/awic/animal-welfare-act.
9. As Rudy (2011) notes, zoos and aquariums typically want to display young and healthy animals because that is what the public has come to expect. As a result, sick, old, uncooperative, or otherwise unmanageable animals are often euthanized, end up in the underground wildlife trade, or, if they're lucky, make it to sanctuaries. During my fieldwork at the Aquarium, I noticed that several animals had disappeared from their exhibit locations. Even after inquiring with several staff willing to chat with me, I was unable to find out where they had gone because the staff I had access to had only heard "rumors."
10. In formal school spaces, the institutional curriculum includes documents like frameworks and standards. While the Aquarium has no such documents, its mission and vision statements, staff handbooks, and funding structure all influence practice. The enacted curriculum, or "what teachers and students actually experience," (Doyle, 1992, p. 493) at the Aquarium includes what staff, visitors, and the animals actually experience.
11. Veronica Pacini-Ketchabaw, Affrica Taylor, and Mindy Blaise (2016) highlight that shifting the focus in educational practice and research from sense-making to learning to be affected "alerts us to being alive and mutually vulnerable to other living creatures" (p. 157).

References

Balcombe, J. (2007). *Pleasurable kingdom.* New York, NY: Macmillan.

Balcombe, J. (2016). *What a fish knows: The inner lives of our underwater cousins.* New York, NY: Scientific American.

Barrett, M. J. (2005). Making [some] sense of feminist poststructuralism in environmental education research and practice. *Canadian Journal of Environmental Education, 10,* 79–93.

Bear, L., Ho, K., Tsing, A., & Yanagisako, S. (2015). Gens: A feminist manifesto for the study of capitalism. Retrieved from https://culanth.org/fieldsights/652-gens-a-feminist-manifesto-for-the-study-of-capitalism.

Bekoff, M. (2007). *The emotional lives of animals*. Novato, CA: New World Library.

Bell, A., & Russell, C. (2000). Beyond human, beyond words: Anthropocentrism, critical pedagogy, and the poststructuralist turn. *Canadian Journal of Education, 25*(3), 188–203.

Berns, G. S., Brooks, A. M., & Spivak, M. (2012). Functional MRI in awake unrestrained dogs. *PLoS ONE, 7*(5), e38027.

Biersack, A., & Greenberg, J. B. (2006). *Reimagining political ecology*. Durham, NC: Duke University Press.

Bradshaw, G. A. (2010). You see me, but do you hear me? The science and sensibility of trans-species dialogue. *Feminism & Psychology, 20*(3), 407–419.

Braverman, I. (2013). *Zooland: The institution of captivity*. Stanford, CA: Stanford University Press.

Chrulew, M. (2011). Managing love and death at the zoo: The biopolitics of endangered species preservation. *Australian Humanities Review, 50*, 137–157.

Despret, V. (2015). We are not so stupid … animals neither. *Angelaki, 20*(2), 153–161.

de Waal, F. (2016). *Are we smart enough to know how smart animals are?* New York, NY: W. W. Norton & Company.

Donovan, J. (2006). Feminism and the treatment of animals: From care to dialogue. *Signs: Journal of Women in Culture and Society, 31*(2), 305–329.

Doyle, W. (1992). Curriculum and pedagogy. In P.W. Jackson (Ed.), *Handbook of research on curriculum* (pp. 486–518). New York, NY: Macmillan Publishing.

Droge, P. (2012). Assessing evidence for animal consciousness: The question of episodic memory. In J. A. Smith & R. W. Mitchell (Eds.), *Experiencing animal minds: An anthology of animal-human encounters* (pp. 231–245). New York, NY: Columbia University Press.

Fawcett, L. (2002). Children's wild animal stories: Questioning inter-species bonds. *Canadian Journal of Environmental Education, 7*(2), 125–139.

Fawcett, L. (2005). Bioregional teaching: How to climb, eat, fall, and learn from porcupines. In L. J. Muzzin & P. Tripp (Eds.), *Teaching as activism: Equity meets environmentalism* (pp. 269–280). Montreal, QC: McGill-Queen's Press.

Fawcett, L. (2013). Three degrees of separation: Accounting for naturecultures in environmental education research. In R. B. Stevenson, M. Brody, J. Dillon, & A. E. J. Wals (Eds.), *The international handbook for research on environmental education* (pp. 409–417). New York, NY: Routledge.

Geertz, C. (1973). *The interpretation of cultures: Selected essays*. New York, NY: Basic Books.

Gough, A. (2013). Researching differently: Generating a gender agenda for research in environmental education. In R. B. Stevenson, M. Brody, J. Dillon, & A. E. J. Wals (Eds.), *The international handbook of research on environmental education* (pp. 375–383). New York, NY: Routledge.

Gough, A., Russell, C., & Whitehouse, H. (2017). Introduction: Moving gender from margin to center in environmental education. *Journal of Environmental Education, 48*(1), 5–9.

Gough, A., & Whitehouse, H. (2018). New vintages and new bottles: The "nature" of environmental education from new material feminist and ecofeminist viewpoints. *Journal of Environmental Education, 49*(4), 336–349.

Hanson, E. (2002). *Animal attractions*. Princeton, NJ: Princeton University Press.

Haraway, D. J. (2008). *When species meet*. Minneapolis, MN: University of Minnesota Press.

Hardt, M. (1999). Affective labor. *Boundary, 26*(2), 89–100.

Hosey, G., & Melfi, V. (2012). Human-animal bonds between zoo professionals and the animals in their care. *Zoo Biology, 31*(1), 13–26.

Latimer, J. (2013). Being alongside: Rethinking relations among different kinds. *Theory, Culture & Society, 30*(7/8), 77–104.

Lloro-Bidart, T. (2015). A political ecology of education in/for the Anthropocene. *Environment and Society: Advances in Research, 6*, 128–148.

Lloro-Bidart, T. (2017a). A feminist posthumanist political ecology of education for theorizing human-animal relations/relationships. *Environmental Education* Research, *23*(1), 111–130.

Lloro-Bidart, T. (2017b). Neoliberal and disciplinary environmentality and 'sustainable seafood' consumption: Storying environmentally responsible action. *Environmental Education Research, 23*(8), 1182–1199.

Lloro-Bidart, T. (2018a). A feminist posthumanist ecopedagogy in/for/with animalscapes. *Journal of Environmental Education, 49*(2), 152–163.

Lloro-Bidart, T. (2018b). A feminist posthumanist multispecies ethnography for educational studies. *Educational Studies, 54*(3), 253–270.

Lloro-Bidart, T. & Russell, C. (2017). Learning science in aquariums and on whale-watching boats: The political deployment of other animals. In M. P. Mueller, D. J. Tippins, & A.J. Stewart (Eds.), *Animals in science education: Ethics, curriculum, and pedagogy* (pp. 41–50). New York: Springer.

Lorimer, J., & Srinivasan, K. (2013). Animal geographies. In N. C. Johnson, R. H. Schein, & J. Winders (Eds.), *The Wiley-Blackwell companion to cultural geography* (pp. 332–342). Malden, MA: John Wiley & Sons.

McKenzie, M. (2005). The 'post-post period' and environmental education research. *Environmental Education Research, 11*(4), 401–412.

Meek, D. (2015a). Taking research with its roots: Restructuring schools in the Brazilian landless workers' movement upon the principles of a political ecology of education. *Journal of Political Ecology, 22*, 410–428.

Meek, D. (2015b). Towards a political ecology of education: The educational politics of scale in southern Pará, Brazil. *Environmental Education Research, 21*(3). 447–459.

Meek, D. (2015c). Learning as territoriality: The political ecology of education in the Brazilian landless workers' movement. *Journal of Peasant Studies, 42*(6), 1179–1200.

Meek, D., & Lloro-Bidart, T. (2017). Introduction: Synthesizing a political ecology of education. *Journal of Environmental Education, 48*(4), 213–225.

Milton, K. (2002). *Loving nature: Towards an ecology of emotion*. New York, NY: Routledge.

Milton, K. (2005). Anthropomorphism or egomorphism? The perception of non-human persons by human ones. In J. Knight (Ed.), *Animals in person: Cultural perspectives on human-animal intimacies* (pp. 255–269). New York, NY: Berg.

Oakley, J., Watson, G., Russell, C., Cutter-Mackenzie, A.,Fawcett, L., Kuhl, G., Russell, J., van der Waal, M., & Warkentin, T. (2010). Animal encounters in environmental education research: Responding to the question of the animal. *Canadian Journal of Environmental Education, 15*, 86–102.

Ogden, L. A., Hall, B., & Tanita, K. (2013). Animals, plants, people, and things: A review of multispecies ethnography. *Environment and Society: Advances in Research, 4*, 5–24.

Pacini-Ketchabaw, V., Taylor, A., & Blaise, M. (2016). Decentering the human in multispecies ethnographies. In C. Taylor & C. Hughes (Eds.), *Posthuman research practices* (pp. 149–167). Houndmills, UK: Palgrave Macmillan.

Pedersen, H. (2004). Schools, speciesism, and hidden curricula: The role of critical pedagogy for humane education futures. *Journal of Futures Studies, 8*(4), 1–14.

Pedersen, H. (2010a). *Animals in schools*. West Lafayette, IN: Purdue University Press.

Pedersen, H. (2010b). Is 'the posthuman' educable? On the convergence of educational philosophy, animal studies, and posthumanist theory. *Discourse: Studies in the Cultural Politics of Education, 31*(2), 237–250.

Rocheleau, D. (2015). A situated view of feminist political ecology from my networks, roots, and territories. In W. Harcourt & I. L. Nelson (Eds.), *Practising feminist political ecologies: Moving beyond the 'green economy'* (pp. 29–66). London, UK: Zed Books.

Rudy, K. (2011). *Loving animals: Toward a new animal advocacy*. Minneapolis, MN: University of Minnesota Press.

Russell, C. (2000). Why study whalewatching? Environmental education, nature experience, and the social construction of nature. In D. Hodson (Ed.), *OISE Papers in STSE (Science-Technology-Society-Environment) Education* (pp. 49–74). Toronto: University of Toronto Press with the Imperial Oil Centre for Studies in Science, Mathematics, and Technology Education.

Russell, C. (2005). "Whoever does not write is written": The role of "nature" in post-post approaches to environmental education research. *Environmental Education Research, 11*(4), 433–443.

Russell, C., & Bell, A. (1996). A politicized ethic of care: Environmental education from an ecofeminist perspective. In K. Warren (Ed.), *Women's voices in experiential education* (pp. 172–181). Dubuque, IA: Kendall Hunt.

Russell, C., & Fawcett, L. (2013). Moving margins in environmental education research. In R. B. Stevenson, M. Brody, J. Dillon, & A. E. J. Wals (Eds.), *The international handbook for research on environmental education* (pp. 369–374). New York, NY: Routledge.

Russell, C., Gough, A., & Whitehouse, H. (2018). Introduction: Gender and environmental education in the time of #MeToo. *Journal of Environmental Education, 49*(4), 273–275.

Russell, C., & Oakley, J. (2016). Engaging the emotional dimensions of environmental education. *Canadian Journal of Environmental Education, 21*, 13–22.

Singh, N. (2013). The affective labor of growing forests and the becoming of environmental subjects: Rethinking environmentality in Odisha, India. *Geoforum, 47*, 189–198.

Smith, M. (2001). The face of nature: Environmental ethics and the boundaries of contemporary social theory. *Current Sociology, 49*(1), 49–65.

Smith, M. (2011). *Against ecological sovereignty*. Minneapolis, MN: The University of Minnesota Press.

Warkentin, T. (2010). Interspecies etiquette: An ethics of paying attention to animals. *Ethics and the Environment, 15*(1), 101–121.

Warkentin, T. (2012). Thinking like a whale. In J. A. Smith & R. W. Mitchell (Eds.), *Experiencing animal minds* (pp. 129–141). New York: Columbia University Press.

Chapter Seven

A Watershed Moment for Zoos and Aquariums? Exploring Feminist Posthumanist Theories and Pedagogies

Introduction

The author of a recent *Time* magazine article entitled, "The Future of Zoos: Challenges Force Zoos to Change in Big Ways" notes, "Across the U.S. and around the world, zoos are finding that balancing the demands of entertainment, education, and conservation is increasingly difficult" (Worland, 2017, para. 6). Reflecting what might be called a watershed moment, zoos and aquariums now seem especially poised to grapple with their colonial and imperial roots (Bender, 2016; Lukasik, 2016), which neoliberal ideologies have helped to repackage through commodification, individualization, and the management of genes, bodies, and populations (e.g., Braverman, 2013; Chrulew, 2011; Rutherford, 2011). As the previous chapters have demonstrated, zoos and aquariums biopolitically regulate and commodify animal bodies in a multiplicity of neoliberal ways—through ticket sales offering the typical zoo-goer experience, through direct access to animal bodies by purchasing premium "animal encounters" reserved for the elite who can afford them, and even through the packaging of the opportunity to feed a wild animal, which only costs a few extra dollars once an admission fee has been paid. These neoliberal zoo and aquarium ideologies also individualize human action by teaching visitors that personal behavioral changes, such as purchasing sustainable seafood and "adopting an animal," are the primary means through which to effectuate positive environmental change, sometimes foreclosing possibilities for collective action. And finally, the biopolitical management of genes, bodies, and populations, where captive animals are on the one hand valued because of the specific arrangements of their DNA molecules and on the other only "count" as populations worth saving, suggests that the individual animal life matters little in the alphabet soup of accounting and tracking procedures taking place in the name of conservation (see Braverman, 2013). Although these neoliberal apparatuses and technologies come with the potential promise that they might "fix" a host of ecological, environmental, and even social problems (Higham & Neves, 2015), in reality they can be quite problematic.

Yet setting aside for a moment the pitfalls of neoliberal ideologies interpenetrating zoos and aquariums, it is important to consider the possibilities of a so-called watershed moment in zoo and aquarium practice brought on by animal welfarists, zoo practitioners who care about and work closely with animals, and even neoliberalism's discontents. A few years ago I attended an animal studies institute that brought together academics and professionals from a range of disciplinary backgrounds, all tackling thorny questions about human-animal relationships in diverse contexts. Performance artists, anthropologists, sociologists, and even several veterinarians surrounded me in the conference room each day. Each of us took turns sharing our work over the course of a week and when it came time for a marine mammal veterinarian, Michelle to speak, affect penetrated the room so thickly you could almost cut it with a knife.[1] My gut had that same uneasy feeling that used to settle there when I sometimes conversed with Jennie, the Aquarium's volunteer animal educator and keeper featured in Chapters Four and Six of this book. While a few participants shared images or music with the group while discussing their work, Michelle's photos and animated descriptions of animals she has cared for affectively, emotionally, and materially engaged us in ways that many of us were not quite prepared to handle.[2] One of her photos depicted, for example, a radiograph of a marine mammal with multiple bullets lodged in his body and head. With a slightly cracking voice, Michelle explained that marine mammal veterinarians employed by aquariums and marine mammal care facilities often have to make split second decisions in order to save injured wild animals and sometimes human animals too. Inevitably, many of the animals are so severely injured they would never be able to survive in the wild, so they must have a safe place to retire. Michelle assured us that zoos and aquariums are typically these safe places for injured animals, while also disclosing how these institutions can sometimes be perilous for the people working there. In their efforts to shut down these institutions, some animal welfarists and activists have bullied and harassed zoo and aquarium professionals, sometimes even making death threats against them (Parker, 2017; Michelle, personal communication, 2017). While I remain exceedingly supportive of and sympathetic to animal abolitionists and welfarists' causes (e.g., Francione & Garner, 2010) and identify as an animal advocate myself, I also believe it is critical to consider the welfare of zoo and aquarium professionals like Michelle and Jennie. Such concern for human welfare alongside animal and environmental justice issues reflects a critically significant intersectional position I take in all of my academic, activist, and community-based work.

In this concluding chapter, therefore, I begin by drawing on Chapters Three through Six to tease out how the political ecology of education approach taken in each illuminates the ways in which particular aspects of animal-focused education are deeply influenced by neoliberal and biopolitical ideologies. I review how the Aquarium unwittingly structures individualist subjectivities that define human-animal relationships as consumerist and commodified, how animal bodies are biopolitically managed and regulated in the service of edutainment, and how feminist posthumanist perspectives offer a challenge to neoliberal practices. In the latter half of the chapter, I draw on various strands of research in animal-focused education to outline two distinct possibilities for re-thinking zoo and aquarium practices that I hope will challenge, instead of reify, neoliberal ideologies. In so doing, I avoid universally indicting people, institutions, and places as "anti-animal" and instead work to create a space where we might look forward to and make changes together. Nearly all of the 40 Aquarium professionals I interviewed, observed, and conversed with over the course of 14 months of fieldwork demonstrated a deep passion for helping animals and/or the environment and they genuinely believed that the Aquarium provided them a place to embody that passion by improving the lives of animals there. Fewer, however, were critically reflective about their own educational practices or those the institution embraces, and none ever discussed the colonial or imperial histories of zoos and aquariums.[3] As well, animal captivity emerged as a rather mundane and normal part of daily life, suggesting that a significantly more critical approach to education in zoos and aquariums is needed if a truly watershed moment is to actually transpire.

Political Ecologies of Education: Neoliberalism and Animal Focused-Education

By invoking various political ecology of education frameworks in this book, I have addressed the under-theorized and political aspects of teaching and learning in zoos and aquariums, particularly those that enroll animals in edutainment endeavors designed to foster specific kinds of environmental citizenship tied to markets. Chapter Three, for example, illuminated how the Aquarium draws on disciplinary and neoliberal environmentalities to visibilize a problem (declining fish populations and seafood stocks), define instruments and procedures to solve that problem (sustainable seafood consumption), and to pre-suppose the formation of identities that embrace the solutions to this problem ("enlightened" consumers who know how to make informed purchasing decisions to benefit animals or the environment). Chapters Four and Five explored how the Aquarium biopolitically manages animal bodies in order to maintain its own fiscal survivability and to promote a very specific sort of conservation education. In the following sections, I summarize the major points addressed in each of these chapters, teasing out the potential implications for environmental education research, as well as zoo and aquarium practice.

Structuring Commodified and Consumerist Human-Animal Relationships

The forms of environmental citizenship the Aquarium promotes through its governmentalities, as Chapter Three highlighted, are not without flaws. The discourses at the Blue Cavern exhibit, for example, work in tandem with the Seafood for the Future Program to attempt to inspire visitors to care about declining animal populations in the ocean, particularly those serving as human food sources. Yet a hidden curriculum also communicates that fish, mollusks, and crustaceans have value primarily because they are a food source. Further, the solution to declining fish and seafood "stocks" centers on leveraging the market to effectuate positive change—in this case a continual supply of food provided through a complex labeling system that works to ensure that people only consume those types of seafood deemed sustainable. In the context of the Blue Cavern show, a political ecology lens sheds light on how "sustainability" seems to operate at several different scales. First, it is defined by what conservationists classify as judicious and thoughtful fishing practices that enable populations of animals to replenish themselves adequately, which requires people who fish to embrace certain practices in harvesting or "production." Second, at the local scale, businesses must be keen to embrace sustainability by purchasing seafood from fisher folk (or corporations) who have fished judiciously and thoughtfully, so that they can offer sustainable seafood products to consumers. And again at the local scale, consumers must act sustainably by purchasing these products. In this way, animals become products of sustainable human practice wherein they are defined not by their individual or inherent value, but instead by the viability of their populations and, ultimately, the sustainability of the market. When the Aquarium teaches visitors about declining animal populations in the ocean, however, these scales of sustainability are in some ways flattened as visitors learn that they can "help" simply by being better consumers, thereby narrowing what gets to count as sustainability.

Setting aside the problems associated with eco-labeling systems reviewed in Chapter Three that include but are not limited to a lack of consumer knowledge, financial barriers for low-income populations who cannot afford such eco-labeled products, and the inability of such schemes to effectuate certain kinds of policy changes (Elgert, 2012; Gutierrez & Thornton, 2014; Guthman, 2007), the Aquarium's strategy is fraught for other reasons. Not only does Seafood for the Future, and the concomitant pedagogical discourses at the Blue Cavern show, narrowly frame relationships with some animals as consumerist and commodified, but they also ignore social justice issues connected to seafood consumption. In the summer of 2017 the Aquarium's Seafood for the Future website, for example, featured testimonials from fisher folk in the Gulf of Mexico in a section called "Storied Seafood," which resonates with the "storytelling approach" taken at the Blue Cavern show during my fieldwork.[4] These stories briefly chronicle how fisher folk are working with several national and international organizations to embrace sustainable fishing practices that also do not harm vaquitas, small, critically endangered porpoises found only in the northern Gulf of California in Mexico.[5] However, the testimonials tell us little about the lives of these people involved in fishing practice (Aquarium of the Pacific, 2017). Are they, as workers, treated fairly? Do they and their families suffer from food insecurity issues, particularly given that the fish they harvest are destined for export to satisfy "ethical consumers" in the global north (Barnett, Cloke, Clarke, & Malpass, 2005) who want to use financial means to do something about declining fish populations?[6] Are they responsible for the decline of vaquitas or are there larger political economic and geopolitical factors also playing a role?

In general, my research revealed that the Aquarium's primary focus is conservation education; as a result the institution largely eschews social justice problems in favor of "environmental problems," as if the two should or could ever be siloed as such. Decades of critical scholarship in environmental education, for example, has demonstrated that making connections between social, environmental, and animal justice is important; yet the field of conservation education has generally failed to engage with these critical perspectives (e.g., Kahn & Humes, 2009; Lupinacci & Happel-Parkins, 2015; Maina-Okori, Koushik, & Wilson, 2018; Russell, 2005). Although another complex eco-labeling scheme, Fair Trade is also not without flaws (e.g., Jaffee, 2014) it appears that Fair Trade has at least created a labeling process that takes into account the stability of fish populations, biodiversity and ecosystem stability, as well as workers' rights and community development (Fair Trade USA, 2017).

Biopolitical Management in the Service of Edutainment

Aquarium staff, as illustrated in this book, clearly care about aquarium animals, animals more generally, and what they refer to as "conservation." The vast majority of staff, for example, volunteer their time there, while paid staff tend to be underpaid when taking into consider the level of education and experience they have. Staff members' ontological and epistemological commitments—one can learn to "know" truths about animals and their own relationships with the natural world through tactual and visual experiences with the captive bodies of animals—drive many of their teaching practices. Yet biopolitical thinking indelibly fuels these interactions. From a disciplinary biopolitical perspective (Fletcher, 2010), the Aquarium, like many AZA member institutions, manages animal bodies to make live (Braverman, 2013). This making live not only involves the regulation of animal bodies at the Aquarium to ensure that the institution's captive populations remain healthy, vital, and in some cases capable of breeding, but also entails making live through educational means. As Chapters Three and Five demonstrated, one of the institution's goals is to sanitize some animal bodies in such a way that visitors will come to care for them—and eventually their counterpart populations in the wild—and thus adopt more conservation-minded attitudes and behaviors. While seemingly a non-anthropocentric ethical endeavor—i.e., promoting care and concern for captive animals and the animals in the wild whom they represent—the actual encounter is anthropocentric because visitors are not asked to "become with," "learn with" or "live with" (Yusoff, 2011) the individuals with whom they interact and the animals are still being used and have their lives severely restricted. Instead visitors consume what is contrived to appeal to humans through a sort of biopolitical speciesism, akin though not identical to Foucault's racism, as highlighted in Chapter Five (Foucault, 2008).

From a neoliberal biopolitical perspective (Fletcher, 2010), the Aquarium also manages animal (and human bodies) in order to make live. Yet in this case making live is less tied to an ethical project calling for the flourishing of animal life, but instead is tethered to economics. As James Carrier (2012), Randy Malamud (2017), and Kathy Rudy (2011) have emphasized, most institutions like zoos and aquariums are rarely adequately funded through public means and thus must act like private enterprises in neoliberal capitalist markets, despite the fact that most are not-for-profit institutions. The political ecology analysis in this book reveals that this problem appears especially acute for the Aquarium both because of its unique funding structure indebting the institution to "program services" for long-term survival *and* because of its unique location within Southern California's competitive entertainment markets. When the talking animals visitors have come to know in Disney films or the Peanuts' beloved beagle, Snoopy, are both relatively short distances away at Disneyland in Anaheim and Knott's Berry Farm in Buena Park, respectively, it is no wonder that the Aquarium has felt compelled to adopt an interpretation model that at least on some levels embraces entertainment. However, Disneyland, Knott's Berry Farm, and other theme parks obviously differ from the Aquarium in that they do not currently display captive wild animals nor offer visitors the opportunity to tactually interact with some of them.

As the previous chapters have demonstrated, these tactual encounters have many problematic dimensions. Aquarium visitors might interact with individual lorikeets whom they feed, but ultimately visitors know nothing about their individual lives and experiences, including how their own presence actually interrupts lorikeet lives in complex and sometimes detrimental ways. Similarly, the lived experiences of the small sharks in the pools are muted, and indeed contrived to provide the illusion that Aquarium sharks lead great lives because they dine on restaurant-quality, sustainably caught seafood and enjoy socializing with each other and humans in the confines of the exhibit. Although "safety" is emphasized through the "two finger" touch rule, safety is only one aspect of animal wellbeing; other aspects of shark wellbeing, such as reproductive rights, the right to live in open spaces, or to hunt or forage, are silenced in the name of edutainment.

While lorikeet nectar serves as a biopolitical edutainment tool because it fosters interaction, for the small sharks it is the actual exhibit that biopolitically regulates shark bodies in the service of edutainment. By designing a structure that allows visitors to approach from virtually any side, the Aquarium can almost guarantee that humans who desire a tactual experience with the shark will have one. In contrast, when I visited the New England Aquarium in Boston, Massachusetts in 2013, I noticed that the Shark and Ray Touch Exhibit there only allowed visitors to approach from one wall. Thus, small sharks or rays not interested in interacting with people could easily swim away from the wall and enjoy relatively large amounts of open space. While the New England Aquarium similarly appears to emphasize entertainment, as their "Current Exhibits" website displays the headline, "Visit the New England Aquarium for an entertaining and educational experience!", the exhibit structure itself at least privileges small shark and ray experiences to the degree that the animals can easily escape human touch should they wish (New England Aquarium, 2019).

At the Aquarium the exhibit disciplinarily makes live by attempting to inspire visitors to adopt what are perceived to be ethically minded behaviors, especially toward sharks, but also neoliberally makes live insofar as human-shark interactions (which captive small sharks can only minimally avoid) help to ensure the fiscal survivability of the institution. The biopolitically regulated bodies of these animals, therefore, serve as means toward various human ends, but are not ends in themselves (Smith, 2001; van Buren, 2014). Although research critical of zoos has highlighted many of these issues (Beardsworth & Bryman, Chrulew, 2011), in contemporary times neoliberalism has repackaged these "means," invisibilizing animal experiences in new ways, while also reinscribing forms of environmental citizenship intimately tied to the markets of late capitalism.

Political Ecologies of Education: Challenging Neoliberalism through Feminist Posthumanist Theories of Affect and Care

In Chapters Four and Five, I demonstrated how the Aquarium biopolitically regulates animal bodies through disciplinary and neoliberal means in order to make these animals fit for display, as well as to edutain Aquarium visitors. I also highlighted how these sanitized encounters with captive wildlife are unlikely to invoke care and compassion for animals or the environment. In Chapter Six, in contrast, I showed how one husbandry staff member, Jennie, developed affective relationships with the birds whom she cared for as she had embodied and emotional experiences with them.

In her theories of egomorphism and person-based identification, which I briefly discussed in Chapter Six, Kay Milton (2002, 2005) argues that as long as nonhuman animals *appear* to share mutual understandings with us, our own sense of personhood and their sense of personhood can be developed as humans come to see nonhuman others as "like me." This seems to suggest that humans may form meaningful relationships with nonhumans irrespective of the environment in which the interaction occurs. Yet these theories seem to counter some research in environmental education that clearly demonstrates how not all interactions with nonhuman animals or nature inspire people to see nonhuman animals or nature as "persons." In some cases, a problematic hidden curriculum in fact communicates the opposite, particularly when animals are presented to human learners as objects rather than as subjects (e.g., Andrzejewski, Pedersen, & Wicklund, 2009; Fletcher, 2017; Lloro-Bidart & Russell, 2017; Pedersen, 2004). Further, researchers have demonstrated that care and compassion can be quite limited, especially for uncharismatic beings like insects, reptiles, and some fish and birds, as well as for food animals (Bell & Russell, 1999; Boileau & Russell, 2018; Fawcett, 2014; Oakley, 2009; Rice, 2017; Pitt, 2018; Small, 2012). Following this line of reasoning, I contend that development of personhood with nonhumans is impacted by the *quality* of experiences people have with the nonhuman world, as well as the kinds of beings with whom they interact (e.g., charismatic versus uncharismatic species).

Aquarium staff members, particularly those who work in the Husbandry Department like Jennie, have regular and sustained contact with animals. As demonstrated in Chapter Six, they remain highly involved in animal care and labor both materially and affectively (Hardt, 1999; Hardt & Negri, 2000, Singh, 2013), sometimes developing close relationships with the Aquarium animals. Even Rose, the Education staff member in the Lorikeet Forest who was repeatedly bitten by a lorikeet, as described in Chapter Four, developed empathic, emotional, and intersubjective relationships with lorikeets despite exposure to their occasional brutality and despite not viewing their actions as "like me." In contrast to Jennie, Rose grounded her intersubjectivity with the lorikeets in a lack of physical interaction, whereby she and the birds could live mutually undisturbed. This lack of interaction enabled respect for them as living beings not "like me," but as romanticized wild Others. In contrast, Jennie, as discussed in Chapter Six, derived her intersubjectivity from affective and sustained relationships with the lorikeets where she sees them as "like me," despite the lorikeets regularly defecating on her and biting her. For Jennie, the lorikeets are "like me" because of both their "positive" characteristics (they "love" each other, they are identifiable and have names, they feel pain) and "negative" characteristics (they bite and defecate). Despite Rose romanticizing the birds as wild Others, her case suggests that intersubjectivity need not be not predicated on personhood and that violence can preclude a human from seeing an animal as "like me." Jennie's case intimates the opposite—her intersubjective relationship with the lorikeets derives from her ability to view them as persons even though she was regularly defecated on and bitten.

Like Jennie, many staff working directly with the birds in the Lorikeet Forest, do develop ongoing relationships with them. One afternoon there, Justin, another Education and Husbandry staff member, brought me over to where a few Husbandry folks interacted with birds near the Hut. Below is a field note excerpt from that day:

> *At the end of my time in there, Justin took me over to where some Husbandry staff were interacting with the birds. One of them (Cindy) was allowing one of the birds to kiss her on her mouth. It was obvious she genuinely cared for the little green bird. The look in her eyes as she glimpsed its face told a story about how her relationships with the birds were not sanitized theme-park interactions, but rather empathic (think "intersubjective") understandings of another creature. She then went right out to where some of the birds were perched on the trees and put them on her finger. I was a little shocked to see these kinds of body-to-body contact, given that there were many visitors present in the Forest and visitors, of course, are prohibited from having these*

> *kinds of interactions with the birds. Having been savagely bitten by a pet parrot years back myself, and knowing that my husband got stitches in his face after his pet parrot ripped his lip open, I told her she was brave for letting the bird kiss her face. She snapped back that she just knows which ones she can allow to do that, indicating to me that her relationships with the birds involved more than just providing the necessary nutrition and shelter.*

Just a few moments later, I overheard Justin talk to a group of boys that all looked about five years old and appeared to be roughhousing:

> *Justin (to the boys): What is going to happen when you scare the birds? You think they have a sharp beak? You see how sharp their beaks are? You see these cuts and welts left on my hand? It's all from the birds. It's not a good idea. So, how about we don't try to scare the birds? It's not a good idea. Thank you.*

Juxtaposing these different interactions reveals striking differences. While the Husbandry staff member allows the birds to kiss her face and has them perch on her finger, and while Jennie labors materially and affectively for them several days a week since she is entrusted with their care, the boys in the excerpt above only have limited contact with them. Indeed, none of the visitors I interviewed in my study discussed developing affective relationships with Aquarium animals that compared to those of staff members. In this way, the institution's need to serve large numbers of visitors through highly regulated animal-human interactions limits the range of affective and emotional experiences visitors can have with Aquarium animals. In the next section of this chapter, then, I explore what other possibilities exist for human-animals relationships in aquariums and zoos.

Re-Imagining a Feminist Posthumanist Animal-Focused Education for Zoos and Aquariums

In this concluding section, I remain hesitant to generate a normative argument that calls for the immediate abolishment of all zoos and aquariums because I think this would be detrimental to the animals currently residing there and to the thousands of zoo and aquarium staff members who tirelessly provide their care. I am also hesitant to engage with "grand narratives" (Nxumalo & Pacini-Ketchabaw, 2017) that presuppose all zoo and aquarium encounters, interactions, and relationships as inherently "bad." Yet, following the work of political ecologists like Rob Fletcher (2010, 2014, 2015) and John Morgan (2017), as well as critical animal studies scholars in environmental education and beyond who argue that exploitive political economic relationships shape the "animal condition" (Corman & Vandrovcová, 2014; Kahn, 2011; Kahn & Humes, 2009; Pedersen & Stănescu, 2012), it is important to think critically and generatively when considering the realities of life for zoo and aquarium animals, especially given that most are either purposely bred for captivity or are captured from a life in the wild.[7] In this final section, then, I briefly engage select posthumanist and feminist theories (Marran, 2011; Pedersen, 2010; 2013; Russell & Bell, 1996; Yusoff, 2011) to explore two possible future scenarios. In the first scenario, aquariums and zoos continue to operate with most current programs in place, yet they embed the "truths" of captive life in their interpretation schemes to more fully capture individual animal experience (e.g., Russell & Hodson, 2002). In the second scenario, I follow the lead of diverse scholars in suggesting that zoos and aquariums ought to move away from captive breeding programs and programs that entail harvesting animals in the wild and instead embrace a sanctuary model, recognizing that there are currently many sanctuary models in place and that not all are ideal. This would, clearly, involve a radical re-thinking of zoo and aquarium practice so that these institutions would eventually transform into sanctuaries.

Critical Pedagogies of Captivity

Zoos and aquariums capture, breed, and biopolitically regulate animals in the service of conservation and edutainment ends, but the effects of these practices on the actual animals living at these institutions is not widely publicized or included in educational programming, despite sensational films like *Blackfish* that recently drew attention to the plight of captive orcas at Sea World. That is, human visitors to these institutions inflict banal violence on captive wild animals in myriad ways, yet this banal violence, a function of "thoughtlessness, rather than radical will" (Yusoff, 2011, p. 579), typically goes unnoticed. Based on my fieldwork, what visitors do in fact learn is that humans all over the world inflict violence upon *wild* populations of animals, but that they themselves are not necessarily implicated in this violence and that Aquarium animals are thankfully safe. Indeed, in some ways both staff and visitors believe that Aquarium animals really have it made, since they do not have to worry about predators, nor do they have to forage or hunt themselves. In fact, many Aquarium animals consume restaurant-quality seafood that many humans would never be able to afford. It was only after months of fieldwork and many conversations with staff, for example, that I learned how nectar alters lorikeet physiology and that humans entering the Lorikeet Forest with cups of this nectar disrupt lorikeet territorial patterns. Similarly, it was only after countless hours of observation at the Shark Lagoon exhibit, as well as through conversations with staff and visitors, that I realized how constrained the small shark lives really are, as well as how much human tactual interaction they must endure. It was only after a staff member mentioned a "harvest" of animals one day (and after carefully reading ALL of the Aquarium's placards) that I became aware that "harvest" meant capturing wild animals from their habitats to live a life of captivity. And finally, it was only after months of fieldwork that I noticed how some animals just seemed to vanish and I could never find out their fate. All of these practices became apparent to me only after months of fieldwork precisely because the Aquarium's programming explicitly obscures these practices so that visitors do not realize how their own presence and support of these institutions impact the animals.

Resonating with some aspects of ecofeminist scholarship (e.g., Adams & Gruen, 2014; Donovan, 1990; 2006; Donovan & Adams, 2007; Gruen, 2015), including Constance Russell and Anne Bell's (1996) ecofeminist "politicized ethic of care," posthumanist geographer Kathryn Yusoff (2011) argues that we need to *learn with* other beings rather than *learn from* them: "Such *learning with* … hopes for a crossing between communities and the recognition of non-human claims; the right to exist, to have spaces and places made for them that suit their needs and sensibilities" (2011, p. 584). She also contends that *learning with* entails heightened attention to our own practices, including those that might banally inflict violence on other beings. This seemingly minor distinction between "learning from" and "learning with" has rather radical implications for pedagogical practice. Rather than shrouding the realities of life for Aquarium animals in sanitized discourses that appear at the surface as biocentric claims (Marran, 2011), the institution might embrace a more critical pedagogical perspective (Corman & Vandrovcová, 2014; Corman, 2017) and embed this knowledge in all of its programming. In a posthuman pedagogical move, this would center animal life experiences as uncomfortable as that might be for visitors to know and acknowledge, rather than making it incidental to potential human learning gains or even institutional fiscal survivability.

I recognize that embedding knowledge about violence in actual pedagogical practice might reflect a "doom and gloom" educational approach that could disenfranchise some people (Kelsey & Kool, 2012) and strip animals of their subjectivity (Corman, 2017). For example, Corman importantly highlights how pedagogies that only focus on animal suffering or victimhood can be reductive and that it is important to engage learners in a "fuller realization of other animals' lives, including their pleasure" (2017, p. 261). However, at the Aquarium, there is little to no acknowledgement that animals at the Aquarium might suffer. Indeed, visitors are taught that Aquarium animals have a great life—one that is in some ways better than what they would experience in the wild. Thus, in order for visitors to have a fuller realization of other animals' lives, I contend that visitors ought to be able to critically engage with and reflect on their own contributions to animal suffering (e.g., supporting animal captivity). If carefully crafted, this approach could in fact be generative (e.g., Russell & Hodson (2002).

Further, I want to suggest that it is time the zoo and aquarium field move past a focus that privileges the visitor experience and toward that of the animals actually living in captivity. As many critics have highlighted (Beardsworth & Bryman, 2001; Hancocks, 2001; Milstein, 2009), modernized and naturalized exhibits might make some people feel better when they visit zoos and aquariums since these exhibits can create the illusion of animal satisfaction; however these exhibits as well as the ensuing programming, are frequently designed only with humans in mind. While educators certainly do not want to disempower their learners, people should not leave zoos and aquariums feeling empowered at the expense of the animals living there.

Jason Lukasik (2013), for example, highlights how his work at a zoo led him to be involved with marginalized urban high school youth through an outreach program. To design his curriculum and pedagogy, Lukasik describes how he drew on educational theorists like Chet Bowers, Paulo Freire, and John Dewey "to create a space where my students might pursue their own curiosities" (2013, p. 5). Although Lukasik (2013) highlights some of the problems and tensions with his work, such as marginalized students becoming interpreters at the zoo and thereby having the "opportunity to share in the colonial dream—to be empowered at the expense of another creature" (2013, p. 7), he also explores some potential reforms. Rather than "becoming an animal scientist" or "doing citizen science," which are programs operated at zoos and aquariums all over the world, Lukasik (2013) called on his students to "write from the animal perspective," including drawing on insights from animal biology and ecology as well as socio-political issues that might impact the animal. Similarly, Russell (2019) demonstrates how the use of cartoons and graphic novels in "pedagogies of discomfort" can provide students a non-threatening way to engage with disturbing content (also see Lupinacci (2018)).

If carefully crafted, programs like those Lukasik (2013) describes could permit visitors to explore these issues not only in the context of animals living in the wild, but also those living at zoos and aquariums. With such knowledge, visitors might become increasingly dissatisfied with the current zoo and aquarium model and therefore be inspired to effectuate change (e.g., support sanctuaries instead of zoos and aquariums; pressure zoos and aquariums to modify their practices). Further, curricular and program designers could provide opportunities for visitors to deal with any negative emotions they might incur through art or music-based learning modalities (Burkhart, 2016; Leavy, 2009; Sandri, 2013). They also could draw on action-based approaches in EE (e.g., Breunig, Murtell, Russell, & Howard, 2014; Russell & Hodson, 2002) to go beyond individualistic and neoliberal solutions. For example, instead of simply offering sustainable seafood consumption as a solution to overfishing and unsustainable fishing practices, educational programming could provide a list of volunteer opportunities for visitors that include one-day events like beach clean-ups, more sustained programs like volunteering opportunities, as well as learning about and/or involvement in more systemic actions through animal advocacy and animal/environmental policy.

A Watershed Moment? A Shift to Sanctuary Models

The second shift in practice I propose here is more radical, yet perhaps more truly reflective of a watershed moment. As I have demonstrated in this book and as others have highlighted elsewhere (Fletcher, 2014; Rudy, 2011; Rutherford, 2011), late capitalism and neoliberalism have deeply impacted zoo, aquarium, and other ecotourism interpretation practices. Although many of these operations are non-profits, they act as private enterprises due to restrictive funding schemes and the divestment of public funds to maintain costly operations. As a result, these institutions continue to cater to the perceived needs and wants of visitors, sometimes sidelining the needs and wants of animals. Further, AZA member institutions prioritize conservation (Braverman, 2013), allowing captive breeding and even the harvesting of animals from wild spaces. Captive breeding programs continue to serve as a justification for the existence of zoos and aquariums despite evidence that in most cases, such programs fail to successfully re-establish wild populations of animals (Marino, Bradshaw, & Malamud, 2009).

Although the number of existing institutions calling themselves "sanctuaries" is vast and these institutions reflect a great amount of diversity, sanctuary models hold promise because they are typically not indebted to edutainment in the same manner of zoos and aquariums and therefore can prioritize animal welfare over public satisfaction (Marino et al., 2009). Further, sanctuaries typically do not maintain captive breeding programs since the goal is not to preserve specific genes, populations, or species. They also do not typically aim to maintain animals for display. Instead, sanctuaries provide spaces where animals rescued from zoos, aquariums, biomedical research facilities, farms, and other precarious situations can safely live out the rest of their lives, especially given that most cannot be released into the wild.

Sanctuaries also differ from zoos and aquariums in that public education is not typically one of the primary goals (Rudy, 2011) even though the American Sanctuary Association's (ASA) mission statement includes "educating the public" (American Sanctuary Association, 2016). A cursory review of sanctuary websites reveals that many do in fact have some educational programming, which might include tours, intern and volunteer programs, and even craft projects where children can make items for the animals (e.g., A New Hope Animal Sanctuary, 2015; International Exotic Animal Sanctuary, 2011; Wildcat Sanctuary, 2017). Tours, which provide visitors an opportunity to visit a working facility that does not necessarily cater to their edutainment needs, are a potentially critical pedagogical practice. If carefully designed, they could assist the public in understanding why sanctuaries exist as well as how they meet the complex needs of the animals there. Intern and volunteer programs, which presumably offer sustained interaction with animals, also hold promise if one considers how, for example, Jennie's sustained experiences with the Aquarium's birds connected her affectively to them in ways that visitors simply could not access. And finally, the Wildcat Sanctuary offers unique educational programming that allows children to actually construct craft projects to benefit the animals, such as enrichment items, beds, and even dens. If these projects are coupled with knowledge about the individual experiences and lives of animals (taking into consideration the age of children so as to not expose them to unnecessary violence), they might offer youth a meaningful way to engage with and learn with animals, rather than simply learning about them or being entertained by them (e.g., Kelsey & Armstrong, 2012). One can imagine, for example, that a child who simply pets a shark has a qualitatively different experience than one who actually builds something the animal needs.

Conclusion

I recognize that there are many barriers preventing an immediate shift to how zoos and aquariums operate, including to their educational programming, not the least of which is funding. Zoos and aquariums often operate in the mode of private enterprise, not always because this is their preferred method of operation, but simply because they must respond to the demands of late capitalism and neoliberalism as well the divestment of public funding. For sanctuaries to avoid falling prey to similar models, they must obtain funding via other means while maintaining their commitment to critical educational programming that challenges the public's belief that animals exist to serve as human entertainment. Further, the cultural significance of zoos and aquariums is woven into the fabric of many Western societies, which, like the Aquarium, use these institutions to provide the urban/suburban public with access to specific kinds of nature and culture while also bringing billions of dollars into cities as people spend copiously at these institutions and in the surrounding areas. The Aquarium, for example, hosts multiple cultural festivals throughout the year to celebrate Long Beach's diverse inhabitants. At the time of my fieldwork, "Night Dive" was an annual event designed to appeal to potential visitors (adults 20–35 years in age; adults without children) who do not typically patronize the Aquarium (Personal Communication, 2013). Currently described as "an adults-only event with local bands rockin' out right next to the fish, DJs spinning in the Aquarium galleries, inspiring works of art, cocktails, and drinks, and tasty bites from food trucks," Night Dive now occurs monthly at a cost of $19.95 for admission, which does not include drinks or food (Aquarium of the Pacific, 2020). Thus, institutions and municipalities will likely be unwilling to sacrifice the financial benefits they reap at the expense of captive animals unless public pressure demands that they do so.

However, North America could be approaching true watershed moment with respect to how we view animal captivity. In 2016, Governor Jerry Brown of California signed the Orca Protection Safety Act, which bans the breeding of orcas in captivity, as well as the use of orcas in entertainment (but still permits their use in educational shows) (Hugo, 2016). While this policy does not affect the Aquarium because it does hold any cetaceans captive, it does impact neighboring Sea World in San Diego, which was featured in the film *Blackfish.* In more progressive moves, early 2018 the Vancouver Aquarium decided to cease keeping dolphins and whales in captivity due to mounting public dissatisfaction and protest. Park board chair, Stuart McKinnon, released the following statement reflecting why this change was made,

> The public told us they believed the continuing importation and display of these intelligent and sociable mammals was unethical and incompatible with evolving public opinion and we amended our bylaws accordingly. We look forward to working with the Vancouver Aquarium as it intensifies its focus on Ocean Wise research and conservation (Lindsay, 2018, para. 9).

And finally, in the summer of 2019, the Canadian Parliament passed a statute to outlaw the captivity of whales, dolphins, and porpoises in Canada. This bill excludes rescues and sanctuaries, suggesting the Canadian public is changing its views regarding the captivity and display of cetaceans (Parliament of Canada, 2019). The Vancouver Aquarium's reaction to public pressure, as well as ensuing Canadian legislation, also suggest that zoo and aquarium professionals are willing to be responsive to change in order to carve out a way forward for animals and people.

Thus, revisiting Michelle's case, the marine mammal veterinarian I described in the introduction to this chapter, it is important to emphasize here that public forms of participation should challenge the social and political economic structures that support current zoo and aquarium models and not individually target the people who work tirelessly in these institutions to care for animals. For example, although we might disagree with the way the Copenhagen Zoo handled the death of Marius the giraffe—creating a public display whereby zoo staff shot the giraffe and then fed him to one of the zoo's lions—sending death threats to Bengt Holst, the zoo's scientific director and public face of the zoo's euthanization and dissection policies, is not likely to effectuate long-term change (Parker, 2017). Instead the public, as in the case of the Vancouver Aquarium, should protest most forms of captivity, captive breeding programs, and culling practices all of which are byproducts of the contemporary zoo and aquarium model that takes captivity for granted, values certain species and genes above others, and culls animals to maintain space for those having the most value.[8] These protests could take many forms, including withdrawing support of these institutions (i.e., refusing to visit them), supporting sanctuaries in myriad ways (e.g., financial donations, volunteering time to care for animals or develop critical education programming), and lobbying governments to enact better animal welfare legislation. The potential watershed moments for zoos and aquariums discussed in this book, therefore, requires public participation that goes beyond the edutainment of an average 2.5-hour zoo or aquarium visit or the bullying of zoo and aquarium professionals, most of whom care tremendously about animals, and instead embraces critical feminist posthumanist practices like those described in this book.

To conclude, this book contributes to EE research in several important ways. First, through a three-pronged political ecology of education approach, I have demonstrated how teaching and learning about the interrelatedness of animals, the environment, and people ought to be understood as nested within various political contexts that deeply shape practice (e.g., Kuhl, 2018). While institutions are located within specific municipalities and regions, none is immune to wider political-economic forces like neoliberal capitalism, which affect the development of emotional relationships with animals, environmental subjectivities, how solutions to pressing problems are framed, and even how to approach the management of life. Second, animal-focused educators and researchers in EE have mounted a significant challenge to ontologies, theories, methodologies, and methods that relegate animals to the realm of objects, rather than as acting subjects (e.g., Lloro-Bidart & Banschbach, 2019; Russell, 2005; Spannring, 2017). This book contributes to that rich body of scholarship by simultaneously demonstrating how animal lives are constrained in captive contexts, while also capturing moments of animal agency and resistance, which Corman (2017) has emphasized is necessary to understand animals on a deeper level. And finally, while the Aquarium's specific funding structure and location in Southern California's entertainment markets has indelibly affected practice in unique ways, many of the issues the Aquarium faces, such as fiscal woes and questions about best pedagogical practice and exhibit design are perhaps not unique. Although my aim in this book was to resist making normative arguments about good or bad practice, the chapters in this book can serve as sort of framework to engage zoo and aquarium professionals in more progressive forms of practice, even while they operate under very specific institutional constraints.

Notes

1. "Michelle" is a pseudonym.
2. Emotion, following the traditions of Charles Darwin, William James and Silvan Tomkins is generally thought of as our subjective experience of physiological changes, experienced as feelings, whereas affects refer to "the capacity to move and be moved …[in] a more generalised way" (as in the work of Benedict Spinoza and Gilles Deleuze) (Kirby & Wilson, 2011, p. 228). Like Vicki Kirby and Elizabeth Wilson, I do not dichotomously conceptualize emotion as individual and psychologized and affect as disseminated and generalized, but recognize that the human mind is co-constituted with "objects, human and nonhuman alike" in a diffuse, circulatory fashion (2011, p. 229).
3. Indeed Jason Lukasik (2013), a former zoo employee and now an academic, notes that his attempts to create a zoo curriculum that included knowledge of the colonial roots of zoos ultimately led to his termination of employment with the Lincoln Park Zoo in Chicago.
4. See the archived story here: http://www.aquariumofpacific.org/seafoodfuture/vaquita_conservation. Accessed September 17, 2020.
5. I use a gender-neutral term here, "fisher folk" even though all of the people featured on the website are men. According to the Missippipi-Alabama Sea Grant Consortium, about 93.9 percent of fisher folk are men and about 6.1 percent of fishing workers and boat owners are female. See: http://masgc.org/news/article/know-more-about-the-commercial-fishermen-in-the-gulf-of-mexico-states. Accessed September 17, 2020.
6. On fishing and food security, see Edward Allison (2011), Allison, Blake Ratner, Björn Åsgård, Rolf Willmann, Robert Pomeroy, and John Kurien (2011). On neoliberalism, oceans, and property rights, see Becky Mansfield (2004). For human trafficking, forced labor, and working conditions in fisheries see Melissa Marschke and Peter Vandergeest (2016) and Vandergeest, Olivia Tran, and Marschke (2017).
7. I recognize that animal lives in the wild are still quite dangerous and violent. Given the widespread impacts of climate change, deforestation, and habitat degradation, many wild animals live in precarious situations.
8. I acknowledge that some animals need to be temporarily held captive before they can be rehabilitated and released into the wild. Further, some animals might be held captive in perpetuity for various reasons, including permanent injury. These animals should have their needs and wants prioritized over that of people visiting the institutions housing them, as well as the institutions themselves.

References

A New Hope Animal Sanctuary. (2015). Get involved. Retrieved from http://www.anewhopesanctuary.org/get-involved. Accessed September 17, 2020.

Adams, C. J., & Gruen, L. (2014). *Ecofeminism: Feminist intersections with other animals & the earth*. New York, NY: Bloomsbury.

Allison, E. H. (2011). *Aquaculture, fisheries, poverty and food security*. Penang, Malaysia: The WorldFish Center.

Allison, E. H., Ratner, B. D., Åasgård, B., Willmann, R., Pomeroy, R., & Kurien, J. (2012). Rights-based fisheries: From fishing rights to human rights. *Fish and Fisheries, 13*(1), 14–29.

American Sanctuary Association. (2016). About American Sanctuary Association. Retrieved from http://www.asaanimalsanctuaries.org/about_ASA.htm. Accessed September 17, 2020.

Andrzejewski, J., Pedersen, H., & Wicklund, F. (2009). Interspecies education for humans, animals, and the earth. In J. Andrzejewski, M.P. Baltodano, & L. Symcox (Eds.), *Social justice, peace, and environmental education* (pp. 136–154). New York, NY: Routledge.

Aquarium of the Pacific. (2017). Storied seafood. Retrieved from http://www.aquariumofpacific.org/seafoodfuture/storiedseafood. Accessed May 9, 2019.

Aquarium of the Pacific. (2020). Night Dive. Retrieved from https://www.aquariumofpacific.org/events/info/night_dive/. Accessed September 17, 2020.

Barnett, C., Cloke, P., Clarke, N., & Malpass, A. (2005). Consuming ethics: Articulating the subjects and spaces of ethical consumption. *Antipode, 37*(1), 23–45.

Beardsworth, A. & Bryman, A. (2001). The wild animal in late modernity: The case of the Disneyization of zoos. *Tourist Studies, 1*(1), 83–104.

Bell, A., & Russell, C. (1999). Life ties: Disrupting anthropocentrism in language arts education. In J. Robertson (Ed.), *Teaching for a tolerant world: Grades K-6: Essays and resources* (pp. 68–89). Urbana, Il: National Council of Teachers of English.

Bender, D. (2016). *The animal game: Searching for wildness at the American zoo*. Cambridge, MA: Harvard University Press.

Boileau, E., & Russell, C. (2018). Insect and human flourishing in early childhood education: Learning and crawling together. In A. Cutter-Mackenzie, K. Malone, & E. Barratt Hacking (Eds.), *Research handbook on childhoodnature: Assemblages of childhood and nature* (pp. 1323–1338). New York, NY: Springer.

Braverman, I. (2013). *Zooland: The Institution of captivity*. Stanford, CA: Stanford University Press.

Breunig, M, Murtell, J., Russell, C., & Howard, R. (2014). The impact of integrated environmental studies programs: Are students motivated to act pro-environmentally? *Environmental Education Research, 20*(3), 372–386.

Burkhart, J. (2016). Singing the spaces: Artful approaches to navigating the emotional landscape in environmental education. *Canadian Journal of Environmental Education, 21*, 72–87.

Carrier, J. G. (2012). Dollars making sense: Understanding nature in capitalism. *Environment and Society: Advances in Research, 3*, 5–18.

Chrulew, M. (2011). Managing love and death at the zoo: The biopolitics of endangered species preservation. *Australian Humanities Review, 50*, 137–157.

Corman, L. (2017). Ideological monkey wrenching: Nonhuman animal politics beyond suffering. In D. Nibert (Ed.), *Animal oppression and capitalism, Volume 2: The oppressive and destructive role of capitalism* (pp. 252–269). Santa Barbara, CA: Praeger Press.

Corman, L., & Vandrovcová, T. (2014). Radical humility: Toward a more holistic critical animal studies pedagogy. In A. Nocella II, J. Sorenson, K. Socha, & A. Matsuoka (Eds.), *Defining critical animal studies: An introduction to an intersectional social justice approach to animal liberation* (pp. 135–157). New York, NY: Peter Lang.

Donovan, J. (1990). Animal rights and feminist theory. *Signs: Journal of Women in Culture and Society, 15*(2), 350–375.

Donovan, J. (2006). Feminism and the treatment of animals: From care to dialogue. *Signs: Journal of Women in Culture and Society* 31(2): 305–329.

Donovan, J., & Adams, C. J. (2007). *The feminist care tradition in animal ethics.* New York, NY: Columbia University Press.

Elgert, L. (2012). Certified discourse? The politics of developing soy certification standards. *Geoforum, 43*, 295–304.

Fair Trade USA. (2017). Product certification: Seafood program. Retrieved from http://www.fairtradeusa.org/certification/producers/seafood. Accessed September 9, 2020.

Fawcett, L. (2014). Kinship imaginaries: Children's stories of wild friendships, fear, and freedom. In G. Marvin & S. McHugh (Eds.), *Routledge handbook of human-animal studies* (pp. 259274). New York, NY: Routledge.

Fletcher, R. (2010). Neoliberal environmentality: Towards a postructuralist political ecology of the conservation debate. *Conservation and Society, 8*(3), 171–181.

Fletcher, R. (2014). *Romancing the wild: Cultural dimensions of ecotourism.* Durham, NC: Duke University Press.

Fletcher, R. (2015). Nature is a nice place to save but I wouldn't want to live there: Environmental education and the ecotourist gaze. *Environmental Education Research, 21*(3), 338–350.

Fletcher, R. (2017). Connection with nature is an oxymoron: A political ecology of "nature-deficit disorder." *Journal of Environmental Education, 48*(4), 226–233.

Foucault, M. (2008). *The birth of biopolitics: Lectures at the College de France 1978–1979.* (M. Senellart, Ed., G. Burchell, Trans.) New York, NY: Palgrave MacMillan.

Francione, G., & Garner R. (2010). *The animal rights debate: Abolition or regulation?* New York, NY: Columbia University Press.

Gruen, L. (2015). *Entangled empathy: An alternative ethic for our relationships with animals.* New York, NY: Lantern Books.

Gutierrez, A., & Thornton, F. (2014). Can consumers understand sustainability through seafood eco-labels? A U.S. and U.K. case study. *Sustainability, 6*, 8195–8217.

Guthman, J. (2007). The Polanyian way? Voluntary food labels as neoliberal governance. *Antipode, 39*(3), 456–478.

Hancocks, D. (2001). *A different nature: The paradoxical world of zoos and their uncertain future*. Berkeley, CA: The University of California Press.

Hardt, M. (1999). Affective labor. *Boundary, 26*(2), 89–100.

Hardt, M., & Negri, A. (2000). *Empire*. Cambridge, MA: Harvard University Press.

Higham, J., & Neves, K. (2015). Whales, tourism and manifold capitalist fixes: New relationships with the driving forces of capitalism. In K. Markwell (Ed.), *Animals in tourism: Understanding diverse relationships* (pp. 109–127). London, UK: Channel View Publications.

Hugo, K. (2016 September 14). Orca shows and breeding banned in California. *National Geographic*. Retrieved from https://news.nationalgeographic.com/2016/09/california-bans-SeaWorld-orca-breeding-entertainment/. Accessed September 17, 2020.

International Exotic Animal Sanctuary. (2011). Education. Retrieved from https://www.wildanimalsanctuarytexas.org/educate. Accessed September 17, 2020.

Jaffee, D. (2014). *Brewing justice: Fair Trade coffee, sustainability, and survival*. Oakland, CA: The University of California Press.

Kahn, R. (2011). Towards an animal standpoint: Vegan education and the epistemology of ignorance. In E. Malewksi & N. Jaramillo (Eds.), *Epistemologies of ignorance in education* (pp. 53–70). Charlotte, NC: Information Age.

Kahn, R. V. & Humes, B. (2009). Marching out from Ultima Thule: Critical counterstories of emancipatory educators working at the intersection of human rights, animal rights, and planetary sustainability. *Canadian Journal of Environmental Education, 14*, 179–195.

Kelsey, E., & Armstrong, C. (2012). Finding hope in a world of environmental catastrophe. In A. E. J. Wals & P. B. C. Corcoran (Eds.), *Learning for sustainability in times of accelerating change* (pp. 187–200). Wagenigen, Netherlands: Wagenigen Academic.

Kirby, V., & Wilson, E. A. (2011). Feminist conversations with Vicki Kirby and Elizabeth A. Wilson. *Feminist Theory, 12*, 227–234.

Kuhl, G. (2018). Sharing a world with wolves: Perspectives of educators working in wolf-focussed education. *Environmental Education Research, 25*(8), 1216–1230.

Leavy, P. (2009). *Method meets art: Arts-based research practices*. New York, NY: Guilford.

Lloro-Bidart, T., & Banschbach, V. (Eds.). (2019). *Animals in environmental education: Interdisciplinary approaches to curriculum and pedagogy*. New York: Palgrave Macmillan.

Lloro-Bidart, T. & Russell, C. (2017). Learning science in aquariums and on whale-watching boats: The political deployment of other animals. In M. P. Mueller, D. J. Tippins, & A. J. Stewart (Eds.), *Animals in science education: Ethics, curriculum, and pedagogy* (pp. 41–50). New York, NY: Springer.

Lukasik, J. (2016, June 7). Is it time to break with the colonial legacy of zoos? *Huffington Post*. Retrieved from http://www.huffingtonpost.com/the-conversation-us/is-it-time-to-break-with_b_10336922.html. Accessed September 17, 2020.

Lukasik, J. M. (2013). Embracing my escape from the zoo: A (sometimes) true account of my curricular inquiry. *Green Theory & Praxis Journal*, *7*(1), 3–16.

Lupinacci, J. (2018). Comics in the classroom: Liberation and everyday superheroes. In J. L. Schatz & S. Parson (Eds.), *Superheroes and critical animals studies: The heroic beasts of total liberation* (pp. 3–17). New York, NY: Lexington Books.

Lupinacci, J., & Happel-Parkins, A. (2015). (Un)learning anthropocentrism: And eco-justice framework for teaching to resist human-supremacy in schools. In S. Rice & A. G. Rud (Eds.), *The education significance of human and non-human interactions: Blurring the species line* (pp. 13–30) New York, NY: Palgrave Macmillan.

Maina-Okori, N. M., Koushik, J. R., & Wilson, A. (2018). Reimagining intersectionality in environmental and sustainability education: A critical literature review. *Journal of Environmental Education*, *49*(4), 286–296.

Malamud, R. (2017). The problem with zoos. In L. Kalof (Ed.), *The Oxford handbook of human-animal studies* (pp. 397–410). New York, NY: Oxford University Press.

Mansfield, B. (2004). Neoliberalism in the oceans: "Rationalization," property rights, and the commons question. *Geoforum*, *35*, 313–326.

Marino, L., Bradshaw, G., & Malamud, R. (2009, March/April). Captivity industry: The reality of zoos and aquariums. *Best Friends Magazine*, 24–27.

Marran, C. L. (2011). Beyond domesticating animal love. *Mechademia*, *6*, 39–50.

Marschke, M., & Vandergeest, P. (2016). Slavery scandals: Unpacking labour challenges and policy responses within the off-shore fisheries sector. *Marine Policy*, *68*, 39–46.

Milton, K. (2002). *Loving nature: Towards an ecology of emotion*. New York, NY: Routledge.

Milton, K. (2005). Anthropomorphism or egomorphism? The perception of non-human persons by human ones. In J. Knight (Ed.), *Animals in person: Cultural perspectives on human-animal intimacies* (pp. 255–269). New York, NY: Berg.

Milstein, T. (2009). "Somethin' tells me it's all happening at the zoo": Discourse, power, and conservationism. *Environmental Communication*, *3*(1), 25–48.

Morgan, J. (2017). Learning from cities: A cautionary note about urban/childhood/nature entanglements. *Environmental Education Research*, *23*(10), 1369–1378l.

New England Aquarium. (2019). Current exhibits. Retrieved from https://www.neaq.org/exhibits/exhibits/. Accessed September 17, 2020.

Nxumalo, F., & Pacini-Ketchabaw, V. (2017). 'Staying with the trouble' in child-insect-educator common worlds. *Environmental Education Research*, *23*(10), 1414–1426.

Oakley, J. (2009). Under the knife: Animal dissection as a contested school science activity. *Journal for Activist Science & Technology Education*, *1*(2), 59–67.

Parker, I. (2017). Killing animals at the zoo. *The New Yorker*. Retrieved from http://www.newyorker.com/magazine/2017/01/16/killing-animals-at-the-zoo. Accessed September 17, 2020.

Pedersen, H. (2004). Schools, speciesism, and hidden curricula: The role of critical pedagogy for humane education futures. *Journal of Futures Studies*, *8*(4), 1–14.

Pedersen, H. (2010). Is 'the posthuman' educable? On the convergence of educational philosophy, animal studies, and posthumanist theory. *Discourse: Studies in the Cultural Politics of Education*, *31*(2), 237–250.

Pedersen, H. (2013). Follow the Judas sheep. Materializing post qualitative methodology in the zooethnographic space. *International Journal of Qualitative Studies in Education, 26*(6), 717–731.

Pedersen, H., & Stănescu, V. (2012). Series editors' introduction: What is "critical" about Animal studies? From the animal "question" to the animal "condition." In K. Socha (Ed.), *Women, destruction, and the avant-garde: A paradigm for animal liberation* (pp. ix-xi). Amsterdam, the Netherlands: Rodopi.

Pitt, H. (2018). Questioning care cultivated through connecting with more-than-human communities. *Social and Cultural Geography, 19*(2), 253–274.

Rice, S. (2017). What's meat got to do with it? Some considerations for ecologizing education with respect to diet. *Educational Theory, 67*(4), 471–489.

Rudy, K. (2011). *Loving animals: Toward a new animal advocacy*. Minneapolis, MN: University of Minnesota Press.

Russell, C. L. (2005). 'Whoever does not write is written': The role of 'nature' in post-post approaches to environmental education research. *Environmental Education Research, 11*(4), 433–443.

Russell, C. (2019). An intersectional approach to teaching and learning about humans and other snimals in rducational contexts. In T. Lloro-Bidart & V. Banschbach (Eds.), *Animals in environmental education: Interdisciplinary approaches to curriculum and pedagogy*. New York: Palgrave Macmillan.

Russell, C., & Bell, A. (1996). A politicized ethic of care: Environmental education from an ecofeminist perspective. In K. Warren (Ed.), *Women's voices in experiential education* (pp. 172–181). Dubuque, IA: Kendall Hunt.

Russell, C. L., & Hodson, D. (2002). Whale watching as critical science education? *Canadian Journal of Science and Technology Education, 2*(4), 485–504.

Rutherford, S. (2011). *Governing the wild: Ecotours of power.* Minneapolis: University of Minnesota Press.

Sandri, O. J. (2013). Exploring the role and value of creativity in education for sustainability. *Environmental Education Research, 19*(6), 765–778.

Singh, N. (2013). The affective labor of growing forests and the becoming of environmental subjects: Rethinking environmentality in Odisha, India. *Geoforum, 47*, 189–198.

Small, E. (2012). The new Noah's Ark: Beautiful and useful species only. Part 2. The chosen species. *Biodiversity, 13*(1), 37–53.

Smith, M. (2001). The face of nature: Environmental ethics and the boundaries of contemporary social theory. *Current Sociology, 49*(1), 49–65.

Spannring, R. (2017). Animals in environmental education research. *Environmental Education Research, 23*(1), 63–74.

van Buren, J. (2014). Environmental hermeneutics deep in the forest. In. F. Clingerman, M. Drenthen, & D. Utsler (Eds.), *Interpreting nature: The emerging field of environmental hermeneutics* (pp. 17–35). New York, NY: Fordham University Press.

Vandergeest, P., Tran, O., & Marschke, M. (2017). Modern day slavery in Thai fisheries: Academic critique, practical action. *Critical Asian Studies, 49*(3), 461–464.

Whale and Dolphin Conservation. (2018 October 25). Bill to outlaw captivity of whales and dolphins finally passes Canadian Senate. Retreived from https://us.whales.org/news/2018/10/bill-to-outlaw-captivity-of-whales-and-dolphins-finally-passes-canadian-senate. Accessed September 8, 2020.

Wildcat Sanctuary (2017). Programs for kids. Retrieved from http://www.wildcatsanctuary.org/education/programs-for-kids/. Accessed September 8, 2020.

Worland, J. (2017, February 15). The future of zoos: Challenges for zoos to change in big ways. *Time*. Retrieved from http://time.com/4672990/the-future-of-zoos/. Accessed September 8, 2020.

Yusoff, K. (2011). Aesthetics of loss: Biodiversity, banal violence, and biotic subjects. *Transactions of the Institute of British Geographers, 37*(4), 578–592.

Author Index

Subject Index

(RE)THINKING ENVIRONMENTAL EDUCATION

GENERAL EDITORS: CONSTANCE RUSSELL & JUSTIN DILLON

The [Re]thinking Environmental Education book series is a response to the international recognition that environmental issues have taken center stage in political and social discourse. Resolution and/or re-evaluation of the many contemporary environmental issues will require a thoughtful, informed, and well-educated citizenry. Quality environmental education does not come easily; it must be grounded in mindful practice and research excellence. This series reflects the highest quality of contemporary scholarship and, as such, is positioned at the leading edge not only of the field of environmental education, but of education generally. There are many approaches to environmental education research and delivery, each grounded in particular contexts and epistemological, ontological and axiological positions, and this series reflects that diversity.

For additional information about this series or for the submission of manuscripts, please contact:

editorial@peterlang.com

To order other books in this series, please contact our Customer Service Department:

peterlang@presswarehouse.com (within the U.S.)
orders@peterlang.com (outside the U.S.)

Or browse by series:

WWW.PETERLANG.COM

www.ingramcontent.com/pod-product-compliance
Ingram Content Group UK Ltd.
Pitfield, Milton Keynes, MK11 3LW, UK
UKHW022207051026
15624UKWH00001B/46

9 781433 147210